Advanced Woven Fabric Design

Advanced Woven Fabric Design

Dr. J. Hayavadana

CRC Press is an imprint of the
Taylor & Francis Group, an **informa** business

WOODHEAD PUBLISHING INDIA PVT LTD

New Delhi, India

First published 2025
by CRC Press
4 Park Square, Milton Park, Abingdon, Oxon, OX14 4RN

and CRC Press
2385 NW Executive Center Drive, Suite 320, Boca Raton FL 33431

CRC Press is an imprint of Informa UK Limited

British Library Cataloguing-in-Publication Data
A catalogue record for this book is available from the British Library

Print edition not for sale in South Asia (India, Sri Lanka, Nepal, Bangladesh, Pakistan or Bhutan)

ISBN13: 9781032840628 (hbk)
ISBN13: 9781032840635 (pbk)
ISBN13: 9781003511021 (ebk)

DOI: 10.1201/9781003511021

Typeset in Times New Roman
by Versatile PreMedia Services

The book is dedicated to

My God
My Parents
My Wife Dr. M. Vanitha
My Son Master Arvind &
My Students, Textile Fraternity

Contents

Preface

I feel elated in dedicating the second book in series for fabric structure titled 'Advanced Woven Fabric Design' to my family, friends world, and textile fraternity. Indeed Advanced Woven Fabric Design is a subject which not only motivates the reader, but also prepares him/her for the practical situations. Normally, a student feels that Advanced Woven Fabric Design as a very tough subject. But in reality, the subject is very simple, interesting, and stimulating. Any reader may feel the heat initially, but once he/she starts studying the advance structures he/she will be more comfortable in further reading.

After teaching the subject from the past three decades, a challenge was always whirling in my mind about writing a book on Advanced Woven Fabric Design with a simple approach and finally I decided to dedicate to textile fraternity which has given me bread and butter. Initially, it was very hard for me to prepare all the designs separately on point paper and then subsequently transfer to a base as I found certain designs very difficult for representation.

I have provided the work sheets and incomplete designs for reader or students for further practice, which is not the case generally. An attempt is also made to describe the loom equipment and construction particulars in respective cases.

Any suggestions in improving the quality and content of the book are most welcome and can be intimated to me so as to improve the readability and accessibility. Lastly, I feel that the book shall fulfil the requirements of reader and will satisfy his/her demands. I thank M/S Woodhead Publishing India in bringing the first edition of this book. I also thank specially Mrs. G.M. Sridevi Kodicherla, freelance graphic designer and computer expert for helping me in typing the text for all chapters without any mistakes.

Prof. Dr. J. Hayavadana

Preface

Foreword

Technology of weaving is the oldest art of producing fabrics since from the time immemorial. Today, the dress code is the index of identification of human kind in the world. Needless to mention here that dress and dress code even it has imbibed by kids specially in selecting their own choice of dress.

The advancements in the world of Science and Technology have witnessed a great deal of changes and weaving is not an exceptional. Weaving include the designing of fabric or in real terms as designing of fabric structure. This subject deals with the structural aspects of woven fabric and its representation along with the necessary loom equipment.

Except one or two books related to weaving or fabric structure, it's hard to find a simple book on fabric structure and design. This book fulfils a requirement which allow a reader to work in the text book itself.

Prof. J. Hayavadana has made a good attempt in reporting the information about complex structures in a lucid manner with number of illustrative examples. The book also give the related weaving or loom equipment required to produce a specific fabric structure.

The book is organized very well. I am very much optimistic that the reader will be benefited with the quality of information provided in each discussion As mentioned above, the salient feature of this book is that a student or reader can work in the text book itself to get the first hand practice of the fabric designs.

Needless to mention here that the publisher M/S Woodhead publishing India has left no way in publishing his other books in a more attractive style with necessary computer setting. I am of the opinion that this book will certainly cater to the needs of the students, research scholars and industrial personnel's. As a matter of fact, one can say that apparel industry will be very much benefited by this book as it deals with the construction particulars and fabric analysis also.

Keeping in view, the need for a comprehensive approach to infuse zeal to learn, this book has been written in a lucid manner by Prof. J. Hayavadana, Department of Technology, University College of Technology (Autonomous), Osmania University, Hyderabad. Who had been in this field for all over two decades.

It is imperative that the understanding of any subject is essentially dependent on the person studying but access to study material should be simple on explanation and act as a guide for full information. The book incorporates both these requirements. The book is divided in to several chapters and in each chapter examples are provided for reader or student for working to understand the design of the structure.

I am sure that this book will be source of knowledge and my overall evaluation of the book is positive. I assure that students, staff, and technologists working in the industry will all find this book very useful.

Prof. Jayanth V. Deshpande
Textile Consultant, Hyderabad.

CHAPTER 1

Introduction to advanced fabric design

We know that the fabric is formed when warp and weft are interlaced. But the question is what is the weight of this fabric if we increase the more series of warp or weft or both threads in fabric formation. Of course, it is to be noted that with an increase in the series of warp and weft, necessary loom arrangements are to be made to produce the desired effects. Hence, when one series of warp and one series of weft are used, we get single-layered structures or simple fabrics. Semi-compound fabrics are those in which more than one type of warp and weft threads are used. For example, in the production of bed ford cord, cutting ends, face ends, and wadding ends are used, and in weft, cutting weft, face weft, backed weft, and wadded weft are also used. Similarly, when more than one series of warp threads are used, compound fabrics or multilayer fabrics are produced. The main purpose of using more series of threads is to increase the weight of the fabrics to attain a specific objective like improving the warmth nature, increasing the weight economically, etc.

1.1 Methods to increase the weight of fabrics

1. **Using extra threads in either warp or weft or in both the directions (Extra Thread Figuring)**

   If extra threads are used in warp direction, the weight will be increased by one and half times and if the disposal ratio is 1:2, the weight will increase by two times and so on. Similar is the concept in weft direction also. Examples of these fabrics are: bed sheets, furnishing fabrics, shirting's, butta fabrics, etc. The loom equipment used will be dobby or jacquard with one or more beams and separate healds for different types of yarns (i.e., ground & extra)

2. **Using an extra thread in the form of Backing thread (Backed Cloths)**

   In this class of fabrics, the weight is increased by either one and half times or two times depending on the RTP. But here, the main purpose of using an extra thread is to increase the warmth nature of fabric economically by using a coarser extra thread as compared to face threads. The backing may be effected either in warp or weft directions. In additions to backing, further the weight is increased by disposing thick threads known as "Wadding " which are not seen to the observer and these may be used in either in warp or weft direction.

3. **Double cloths**

   In this class, two series of warp and two series of weft threads are employed, and hence, the weight will be double the single layer. Basically, the weight increased will be two times the single-layered structure. Weight imparted depends on the RTP of two series of threads, namely face and back. Here also, the weight can be further increased by using wadding threads.

4. **Multilayer fabrics**

   These fabrics include more than two series of warp and two series of weft, and a popular example of this class is treble cloths. In three layers of fabrics, each single layer is formed, giving three times increased weight. These are applied to industrial applications like belting cloth, composite cloth, etc.

5. **Terry Pile fabrics.**

   These are characterized by the presence of loops over the surface. In warp, there will be two series of warp threads and one series of weft. These are extensively used as bath mats, toweling fabrics, etc. These are produced on a dobby provided with a special terry motion. Weight increased is 1.5 times the SLS.

6. **Weft Pile Fabrics**

   Two series of weft and one series of warp constitute the formation of velveteens. These are characterized by the tuft surface. Weight increased is 1.5 times the SLS.

1.2 Figuring with extra threads

1.2.1 Principle of extra thread figuring

A distinguishing feature of these fabrics is that the withdrawal of the extra threads from the cloth leaves a complete ground structure under the figure. The formation of a figure by means of extra threads thus does not detract

Figure 1.1 Cross sections of Extra Weft(a) and Extra Warp(b)

the strength, or weaving quality of the fabric, except so far the extra threads are liable to fray out, whereas in ordinary fabrics, in which the figure is formed by floating the weft or warp threads loosely, the strength of the cloth is reduced somewhat in proportion to the ratio of figure and ground. One of the advantages of figuring with extra threads is that bright colours – in sharp contrast with the ground – may be brought to the surface of the cloth in any desired proportion. Thus, pleasing colour combinations may easily be obtained.

1.2.2 Methods of introducing extra figuring threads

Extra threads may be introduced either as warp or weft, or a combination of both may be employed. For extra warp, the loom must be filled with two beams, and for extra weft, the weaving machine must be capable of introducing more than one type of weft (Fig. 1.1). Extra threads may be introduced in continuous order or intermittent order, depending on the design required to be produced.

1.2.3 Methods of disposing of the surplus extra threads

The disposal of the extra warp or weft threads, in the portions of the cloth where they are not required to form figure, is of great importance. There are many methods in which this can be accomplished:

1. **The extra yarn is allowed to float loosely on the back in the ground of the cloth**

   This method is suitable when the space between two figures is not excessive, when the ground is dense, and when the fabric is used in situations that do not render the long floats objectionable. It is not applicable to clothes in which the ground is so light and transparent that the positions of the extra threads on the back can be perceived from the face side.

2. **The extra thread is allowed to float loosely on the back, and is afterwards cut away**

   This method is mainly suitable for light ground textures, but of the extra threads float somewhat loosely on the surface in forming the ornament, it is necessary for them to be bound in at the edges of the figure, or the loose figuring floats will readily fray out from the surface. The firm interweaving of the extra yarns at the edges, however, makes the outline of the figure less distinct and is rather objectionable unless employed in such a manner as to assist in forming the figure.

3. **By using special threads for stitching the extra threads**

   In compact fabrics, the extra threads are bound in on the underside of the cloth, either between corresponding floats in the ground texture or by means of special stitching threads.

4. **By combining ground with extra floats**

   The extra threads are interwoven on the face of the cloth in the form of small auxiliary figures or floats, thus adding to the fullness of the texture.

1.3 Extra warp figuring

In extra warp figuring, there are two or more series of warp threads to one series of weft threads. The extra threads are introduced by means of second warp beam because of the different take-up rates. The threads can be arranged continuously in the order of 1 extra and 1 ground, 1 extra and 2 ground, 1 extra and 3 ground, and so on. The arrangement may also be intermittent instead of continuous. Where ground area is very large, the extra warp is stitched in sateen order to reduce the length of the extra warp float on the back. Weaves other than sateen can also be used for this purpose. One of the advantages of figuring with extra materials is that bright colours – in sharp contrast with the ground – may be brought to the surface of the cloth in any desired proportion. Pleasing colour combinations, bright or otherwise, may thus be conveniently obtained since the extent of surface allotted to the figuring colour may be readily proportioned in accordance with the degree of its contrast with the ground shade, without the latter being affected.

Figure 1.2 The principle of extra thread figuring for extra warp

1.3.1 Principle of extra warp figuring

The extra threads are allowed to be visible on face or allowed to float at the back or allowed to interlace with ground. The extra threads are mixed with ground in definite proportions and if the extra thread is removed or ravelled from the fabric the original ground remains. This is the fundamental principle of extra thread figuring. Figure 1.2 shows the principle of extra thread figuring.

1.3.2 Spot designs for extra warp

Small spots of figures are frequently employed for ornamenting various types of fabrics (Figs. 1.3 and 1.4). Extra warp in different applications is produced with spot designs varying in shape and size. Spot effects are basically designs chiefly of small, detached spots or figures employed for dress fabrics and other upholstery fabrics. Spotted effects are produced by different ways, such as by employing fancy threads appearing in contrasting colour to occur at intervals and by introducing extra warp or extra weft threads which are arranged on face side of fabrics indicating the spot designs. Ornamental value of the final fabric will improve if warp and weft have colour patterns.

1. By employing at regular intervals fancy threads with spots of contrasting colours in them
2. By introducing extra warp threads worked by dobby or jacquard
3. By extra picks brought to surface where spots are required to form a figure
4. By floating the ordinary weft or warp threads on the surface of the cloth in contrast to the ground weave.

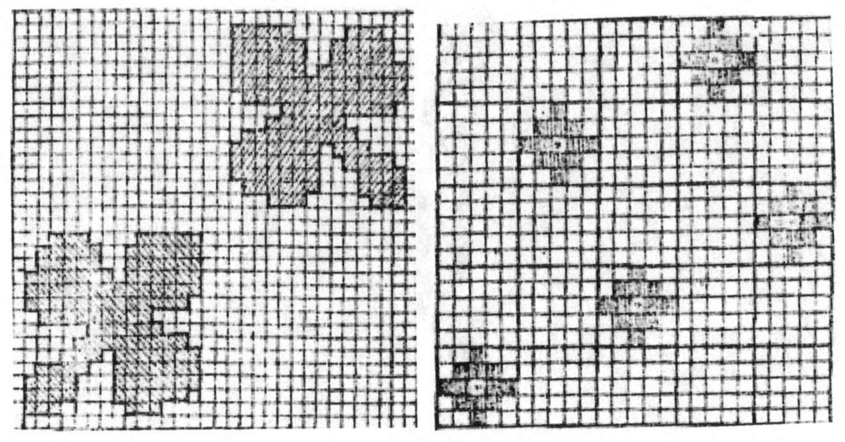

Figure 1.3 Arrangements of Spot effects in fabrics

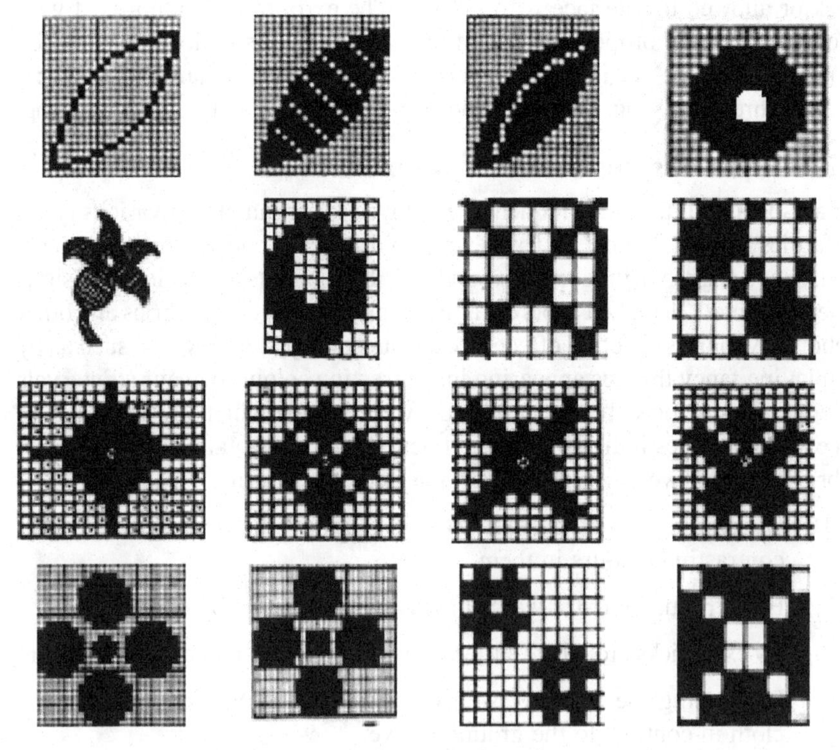

Figure 1.4 Various types of spot designs for extra thread figuring

The spotted designs are produced both in warp and weft directions. However, they are more preferred in weft direction. The reasons are as follows:

1. The weft is normally with less TM (Twist Multiplier) and will be generally more lustrous.
2. The fabric normally contracts in weft direction more in width than length as weft is not under tension as compared to warp and thus weft is brought more prominently on the surface than warp.

The spot figures are sometimes reversed or turnaround in opposite directions to give strong contrasting effects (Fig. 1.3). The spot effects are also arranged on some weave basis like sateen. Figure 1.3 shows such an arrangement where the spot effect is arranged on five-end basis.

As we know, the weft usually forms brighter and clearer spots than the warp. Figures 1.5a, d, e, f show spot designs. The spot design in Figure 1.5a is first developed with extra thread as shown in Figure 1.5b and then with ground as shown in Figure 1.5c. Designs of Figures 1.5d to f are to be developed by reader for cross sections and designs. Further, the reader is also directed to try all the spot designs given in Figure 1.4. In spot designs, we may find either one colour extra warp or more than one colour extra warp depending on the ornamentation requirement. For example, Figure 1.5g indicates a design with 2:1 extra:ground for the spot design shown in Figure 1.5d and Figure 1.5g′ shows the partly developed design for the spot design at Figure 1.5e *wherein the reader is directed to complete the design by drawing the ground.* A spot design in Figure 1.5f is developed with 2:1 in Figure 1.5h and with 1:1 in Figure 1.5i, which is to be completed by the reader with necessary plans. The reader should further note the designs of Figures 1.5j1 to j3 to develop the spot effects along with denting plan. The lifting plan and card punching or pegging are to be prepared by the reader.

1.3.3 Weaving arrangements for extra warp

1.3.3.1 Warping arrangements

Generally, all the advanced fabric use coarser counts such as 10s or 20s. If they are single weaver, it will prefer a sized beam and if they are double, sectional warping will provide the necessary ends along with wide colour combinations in warp.

1.3.3.2 Weft winding arrangements

Semi-automatic or automatic pirn winders are preferred. If the fabric belongs to *Butta* design, then the weaver is given special instructions and the weft is prepared on small pirns on charaka winder manually.

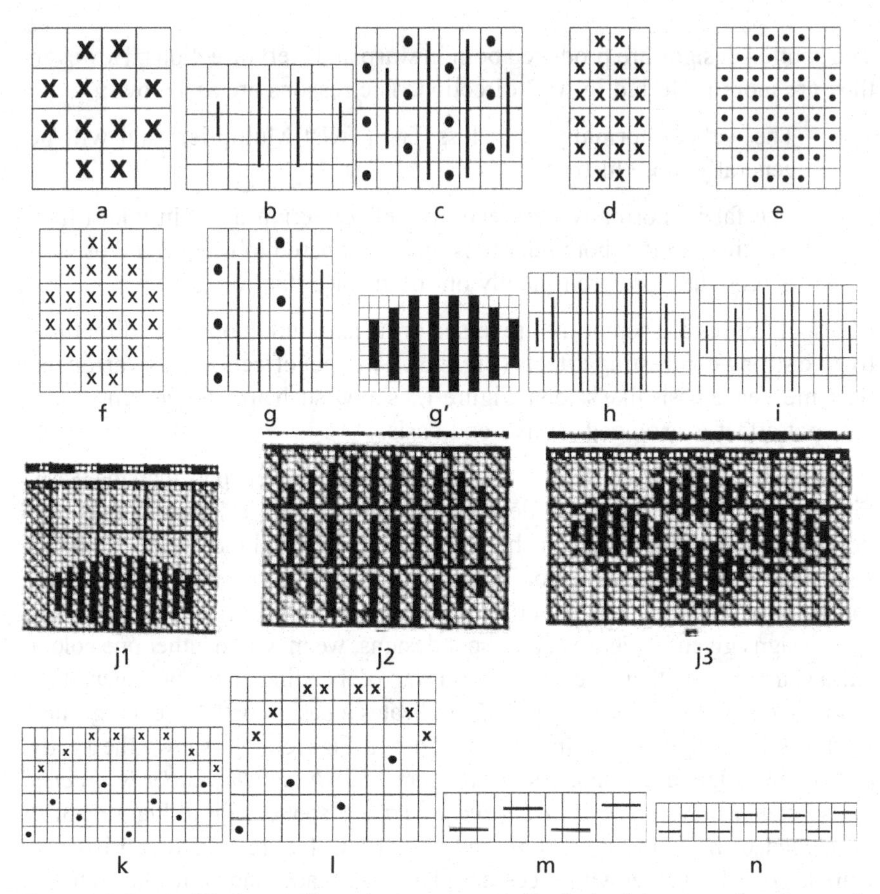

Figure 1.5 (a–j3) Spot effects in extra warp, (k–l) draft in extra figuring and (m–n) denting in extra thread figuring

1.3.4 Draft in extra figuring

Generally, dobby and jacquard looms are used for producing extra thread figuring. It is well known that heald–harness arrangement is the best method for increasing the figuring capacity of shedding device. The arrangement consists of drawing the ground plain on skip draft in the front healds controlled by tappet shedding and extra threads through jacquard harness controlled by dobby or jacquard shedding. Such a set-up is shown in Figures 1.5k and l.

1.3.5 Denting in extra thread figuring

Generally, it is default to draw the RTP (Relative Thread of Proportion) in a dent in advanced structures. Extra threads may be 1:1 or 2:1 or 2:2 and so on,

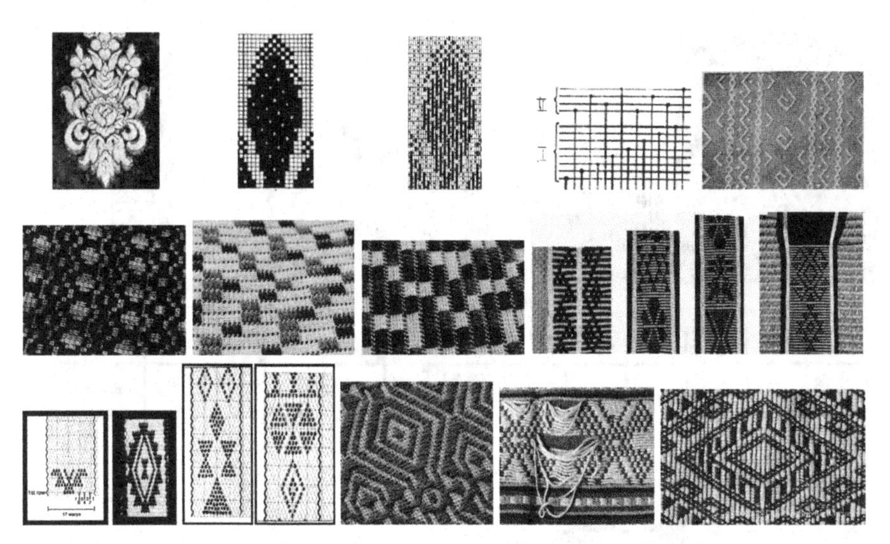

Figure 1.6 Continuous extra warp with weaving drawing set-up.

which are drawn in a dent. Figures 1.5m and n show the ideas. *The reader is directed to draw denting, draft, and peg plans for the designs given above.*

1.3.6 Continuous figuring with one extra warp

In this class of fabrics, the extra warp is allowed to float either on face or back continuously but to bind it the ground pick is passed over it at some intervals so that the end is prevented from fraying out. Figure 1.6 shows a floral motif and a portion is developed at the second stage. The third stage includes developing with ground and binding weave. Here, one should note that the figuring capacity of the jacquard is increased by using heald–harness arrangement as shown in the fourth stage. Here, the first group of ends refers to extra threads and the second group refers to ground ends drawn with skip draft as ground is plain.

1.3.7 Intermittent figuring in one extra warp

Most of the furnishings produced today in decentralized sector account for intermittent variety. Extra thread will be appearing on face as well as back. On face, it may form some figure and on back another figure is formed. One such example is shown in Figure 1.7. The extra warp can be introduced intermittently to form striped or detached figures. Stripes can be produced by floating the extra ends. Continuous or non-continuous extra figures should be placed very close to each other so that the overall design has a striped appearance. Detached figures can be formed on the fabric in one of the bases such as full drop, half drop, Ogee, sateen, and so forth (Fig. 1.7).

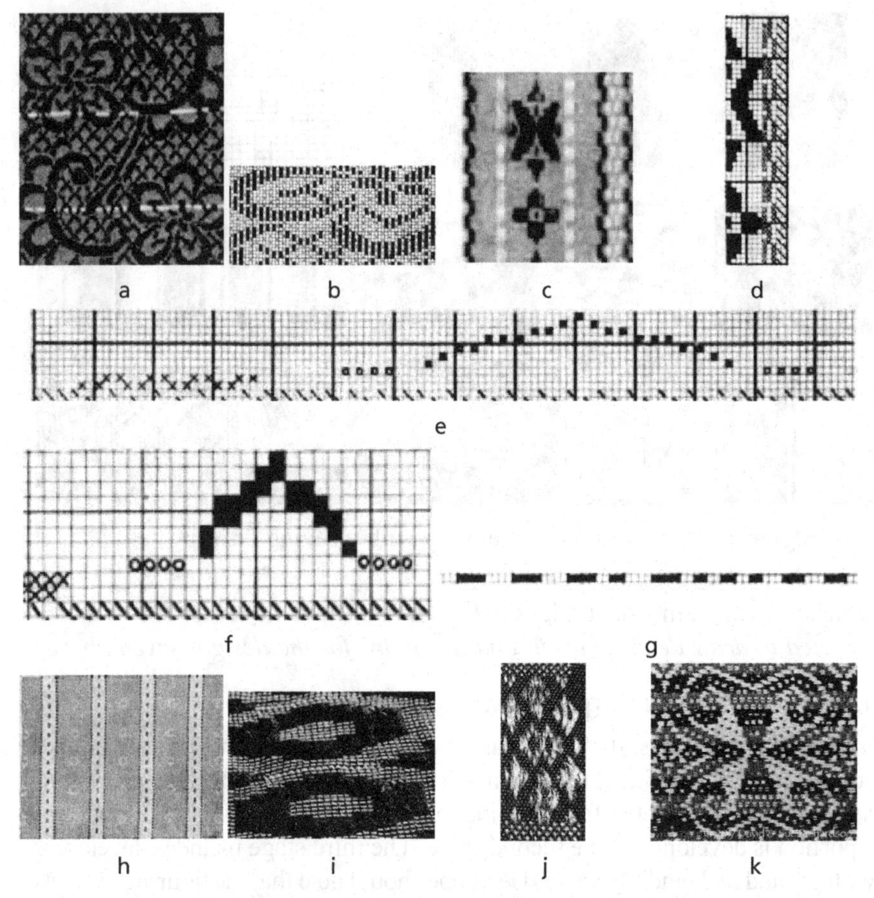

Figure 1.7 Intermittent extra warp fabrics

The Figure 1.7a shows a part of the motif of a furnishing fabric and the design is developed as shown in Figure 1.7b. Another example of intermittent fabric is shown in Figure 1.7c. Figures1.7d to g show the weavers plans for different parts of the motif.

1.3.8 Cut-in effects and clipped spot effects in extra warp

Furnishing fabrics in extra thread figuring are produced on open ground. The extra thread may be allowed to interlace continuously forming figure on face and back or cut away after forming the figure on the back of the fabric by plucking from the surface if it was insecurely anchored to the cloth. The anchorage is achieved in two ways. In one method, the ground ends are dented one per dent and where the extra ends are working they are crammed into a dent on group dent concept along with the ground end. This results in binding of extra ends by ground. In the second method, the extra ends are not permitted to float continuously on the surface but are bound in at regular intervals in such a manner that the discontinuity of the extra end float is not noticeable.

Figure 1.8a shows the cross section to explain the concept of cut-in. Figure 1.8b shows a portion of the design, Figure 1.8c shows the development of a part of the design shown in Figure 1.8b and the second example of cut-in is shown in Figure 1.8e and the developed part of Figure 1.8e is shown in

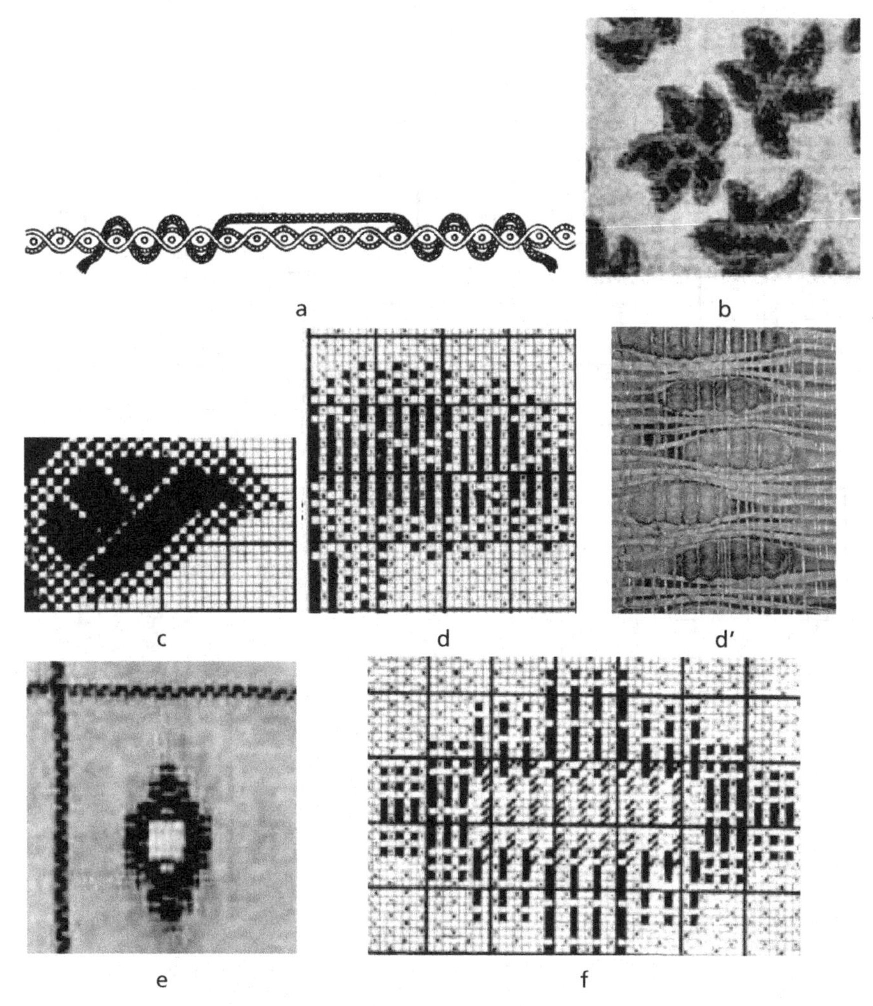

Figure 1.8 (a–f) Cut-in effects in extra warp

Figure 1.8f. Due to a considerable degree of cramming of the extra warp in a single dent and the fashion in which the extra thread is bound gives clipped spot effect in solid appearance. Extra thread figuring is successfully used to produce dobby clipped spot effects with the suitable denting arrangement as mentioned above. These designs are much preferred for furnishing trade. Figure 1.9a shows the denting plan and Figure 1.9b shows one such example

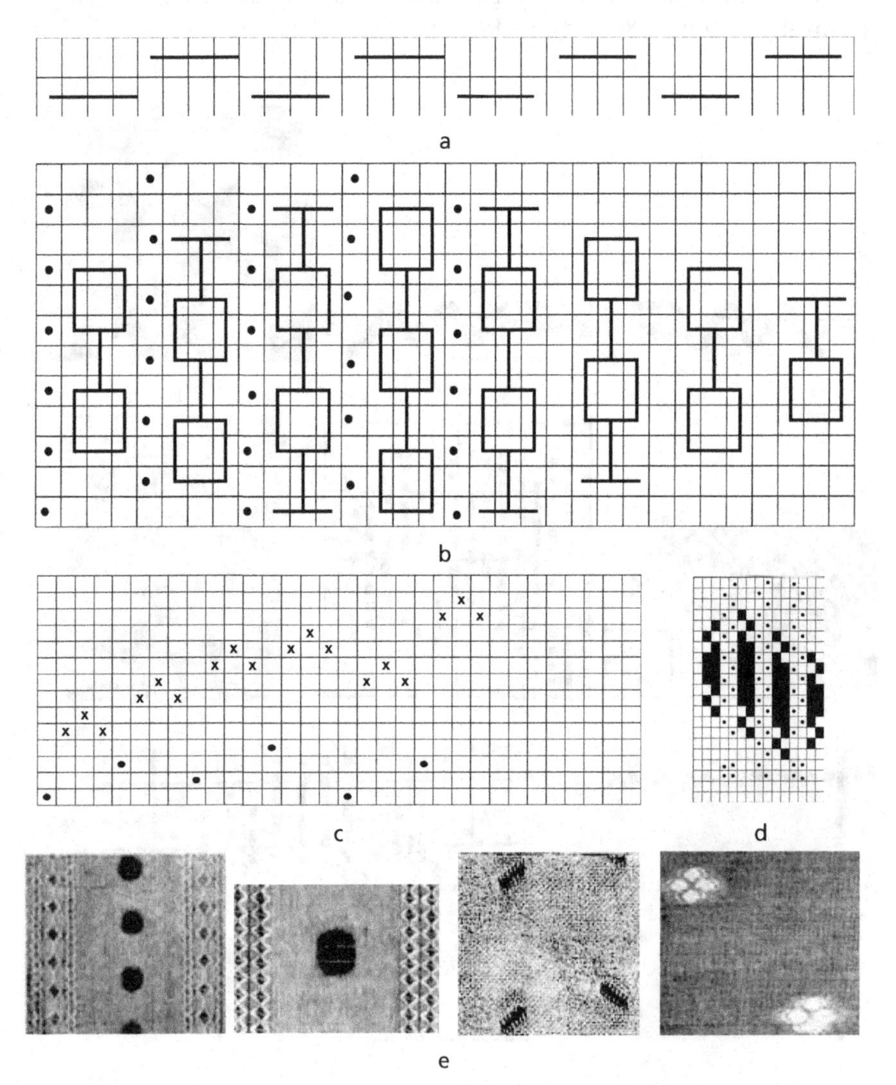

Figure 1.9 (a–e) Clipped spot effects in extra warp

with denting plan in of Figure 1.9a and remaining ground weave is *to be completed by the reader*. The corresponding draft is given in Figure 1.9c for extra figuring and for ground it is to be completed by reader. Figure 1.9e also give the view of the clipped spot fabrics.

Figure 1.9d shows another example of dobby clipped spot effect. The reader is directed to complete the weaver plans.

1.3.9 Extra warp figured effects

Large figured effects can be easily produced in fabrics and following is the example of one such attempt. Figure 1.10 shows a simple design which is suitable for border of different types of fabrics. It can be produced on 73 ends and 156 picks. However, it is to be noted here that the picks are more than warp and thus bumping of reed may occur. Figure 1.10 also shows that a portion of the main figure is developed to form the final repeat. Figure 1.11 shows the two examples of Alhambra Quilt, which is a popular extra warp. The figures show the profile of the motif, development of a part of the profile,

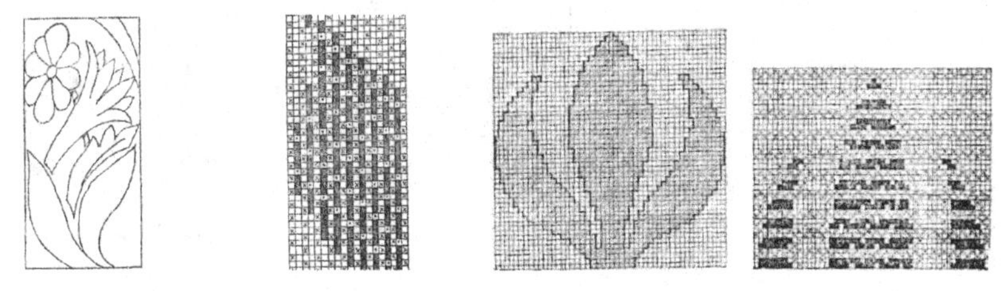

Figure 1.10 Extra warp figured effects

Figure 1.11 Alhambra Quilts

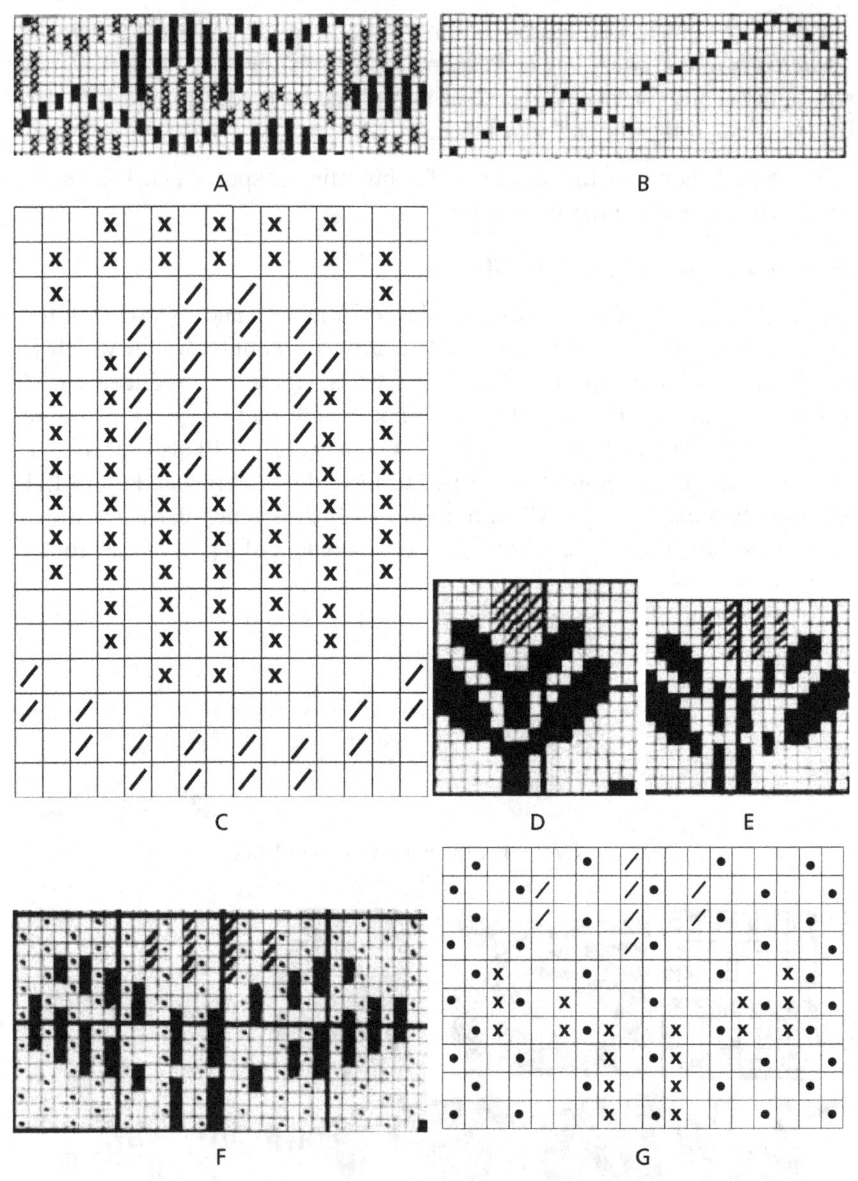

Figure 1.12 (A–G) Two-colour extra warp

and also the final design, which is also a part of the developed design. Further, the reader can also notice the use of decked mail eyes in this production. This special extra warp is woven with two or three ends per decked mail eye where the extra threads are thicker than ground. The suggested quality particulars are as follows:

1. Ground warp: 28 tex, 10 ends/cm, 18–22% crimp
2. Extra thread: 42 tex, group ends per decked mail eye, 11–13 double ends/cm, 7–10% crimp
3. Ground weft: 200 tex, 10 picks/cm

It is to be noted here that in all figured extra warp when the extra thread floats for a longer length, it is customary to bind them along with ground ends or using sateen or twill as ground weave. Further, in some cases, the extra thread is discontinued interwoven with ground to bind the extra thread.

1.3.10 Ornamentation of extra figuring with two coloured warp

Ornamentation can be improved in a basic extra thread figuring by using colour concept. Use of two colours for extra threads with changed RTP is most preferable. Figure 1.12A shows the design, Figure 1.12B shows the draft developed, Figure 1.12C shows the part of the design developed, Figure 1.12D is a second example where two-colour motif is drawn, Figure 1.12E shows the arrangement of two-colour extra warp and Figure 1.12F shows the development of Figure 1.12E and Figure 1.12G shows a slight modification of the design and development thereof. Figure 1.12 shows a style in which two series of extra warp ends are introduced continuously. A feature of the example is that the complete design extends over 50 extra ends, whereas the order of interlacing repeats upon 25 ends. This is due to the figure which has been designed upon an odd number of ends, which causes the colours to change positions in succeeding repeats. The warp colours are also interchanged in the direction of the length of the design.

The usual arrangement of ends in two-colour extra warp ends is 1G:2 extra ends. However, other orders of arrangement of the warp that can also be employed are: 1 ground, 1 extra, 1 ground.

1.3.11 Extra warp planting

'Planting' is a term used for the system of arrangement, which enables a figure to be formed in a large number of colours without an addition being actually made to the series of extra threads (Fig. 1.13).

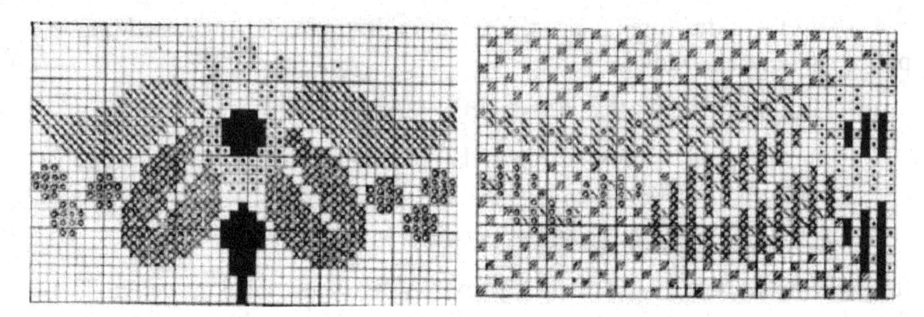

Figure 1.13 Extra warp planting motif

1.4 Extra weft figuring

Extra weft fabrics include two series of wefts and one series of warp. These are called so because extra picks are used in addition to ground picks. Extra weft fabrics may be produced with one or two or more extra weft picks in relation to ground. However, point to be noted is that only one series of warp is used like one series of weft usage in extra warp figuring. The principles of disposal of extra thread in weft are similar to that of extra warp. The effect is obtained by floating the extra weft on face side of ground by interlacing of warp with ground weft at the required places in plain or any other order.

1.4.1 Loom equipment and notation in point paper designs

All types of extra weft figuring production require looms with box motions. Based on the type of extra weft take-up motions are devised or selected. For example, if the selected extra weft is continuous, regular take-up will be useful in calculating the production. However, intermittent take-up is preferred if the weft to be disposed is in intermittent form by rendering the take-up in an operative mode.

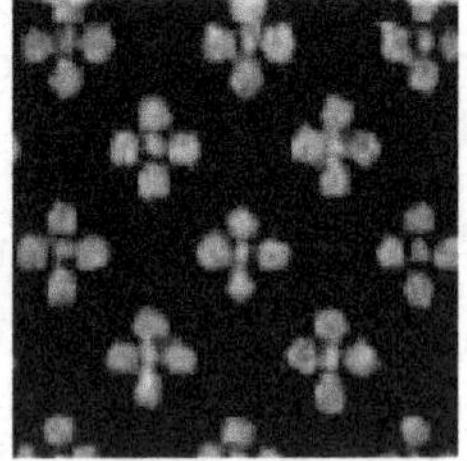

1.4.1.1 Notation

Unlike in majority of the cases where mark indicates warp-up, notations are reversed in extra wefting to facilitate the appearance of design. Hence, the reader is alerted here to make a note of the reversed convention.

1.4.2 Spot effects, continuous figuring with one extra weft

Figures 1.14a & d show the spot design selected for developing extra weft figuring. The figuring is developed in 1:1 ratio for ground:extra weft (Figs. 1.14b, c, & e). 2:1 extra weft:ground is shown in Figure 1.14f. The reader is directed to complete Figures 1.14g to j designs. Figure 1.14c1 shows the spot motif and is developed with ground weave in 1:1 ratio and Figures 1.14c2 and c3 show a developed design for which the spot design has to be traced by the reader. The design in Figure 1.14c4 shows the arrangement of the design of Figure 1.14c1 in 2:2 fashion.

1.4.3 Intermittent effects in extra weft

Extra weft in fabric can be introduced in intermittent way with ground weft in producing detached spot effects. Here, the intermittent extra weft figure is combined with a figure formed by the ground weft. Figure 1.14m shows an example of this class. Generally, the ground weft floats will be close and effectively conceal the binding points, but if the figuring weft is much thicker than ground weft there is a liability of the stitches forcing the ground picks apart and showing on the surface particularly if there is a strong colour contrast between the extra figuring weft and ground. This type of constructions is best suited for producing polyester-type fabric with 160 tex ground and extra weft and 40 picks/cm. Figure 1.14m1 shows the face side of the fabric and Figure 1.14m2 shows the underside of the fabric. Figure 1.14m3 is the developed design of Figure 1.14m1.

1.4.4 Modification of ground weave

It is found that continuation of ground under the figuring is not preferred because ground weft float tends to shield the extra figuring effect and cause the edges of the figure to appear indistinct. Under such circumstances, it is advisable to change or modify the ground weave. (Fig. 1.15a).

1.4.5 Cut-in effects in extra weft

Special cut-in effects are produced in extra weft for exclusive furnishing fabrics in the form of value addition by allowing the extra weft to intervene with ground threads after forming the figure. These effects are also quite popular in the production of what are known as 'Bhutta' designs for ladies blouse material. The extra weft used may be disposed in all over fashion or continuous fashion

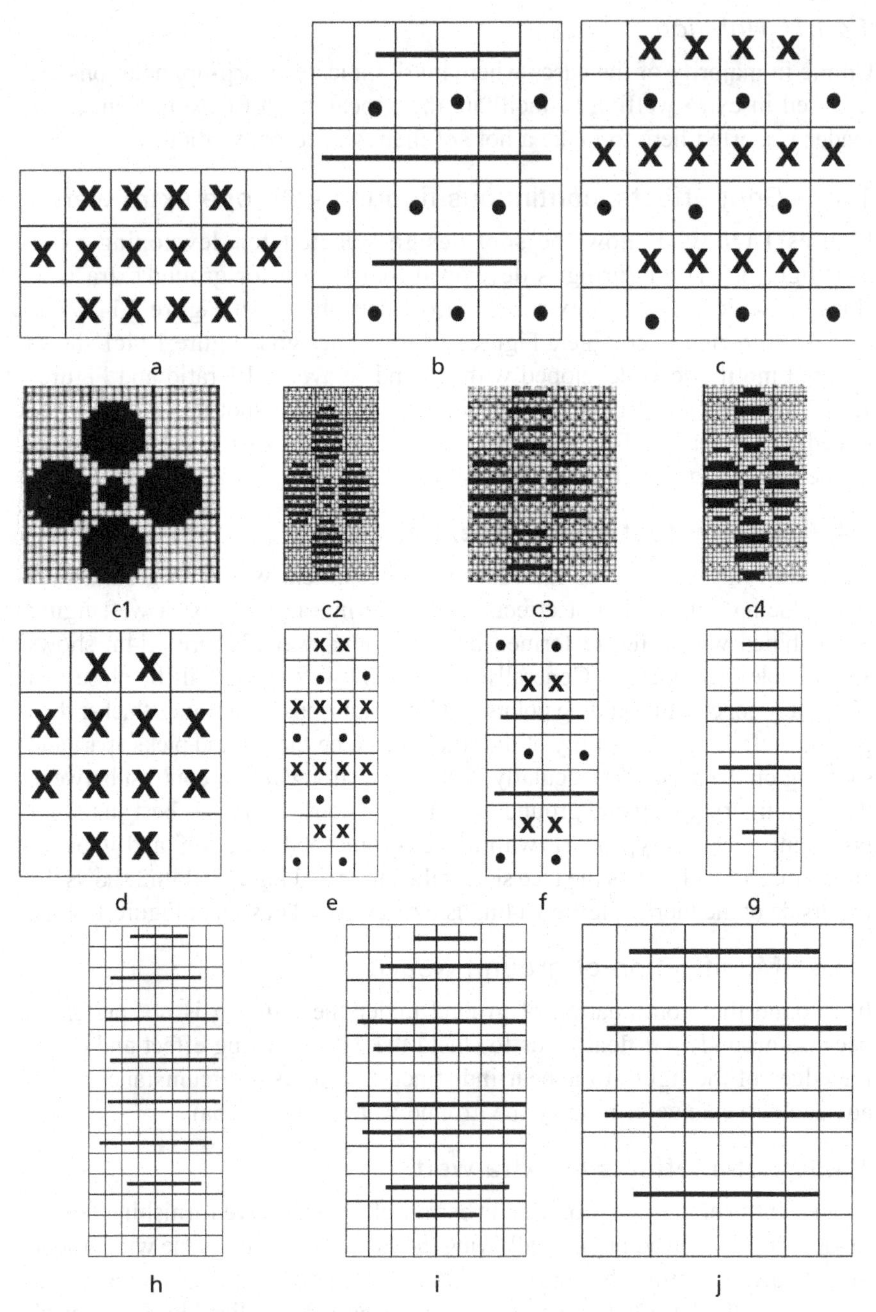

Figure 1.14 (a-m3) Spot effects in Extra Weft and Development of Spot effects (Intermittent effects in extra weft)

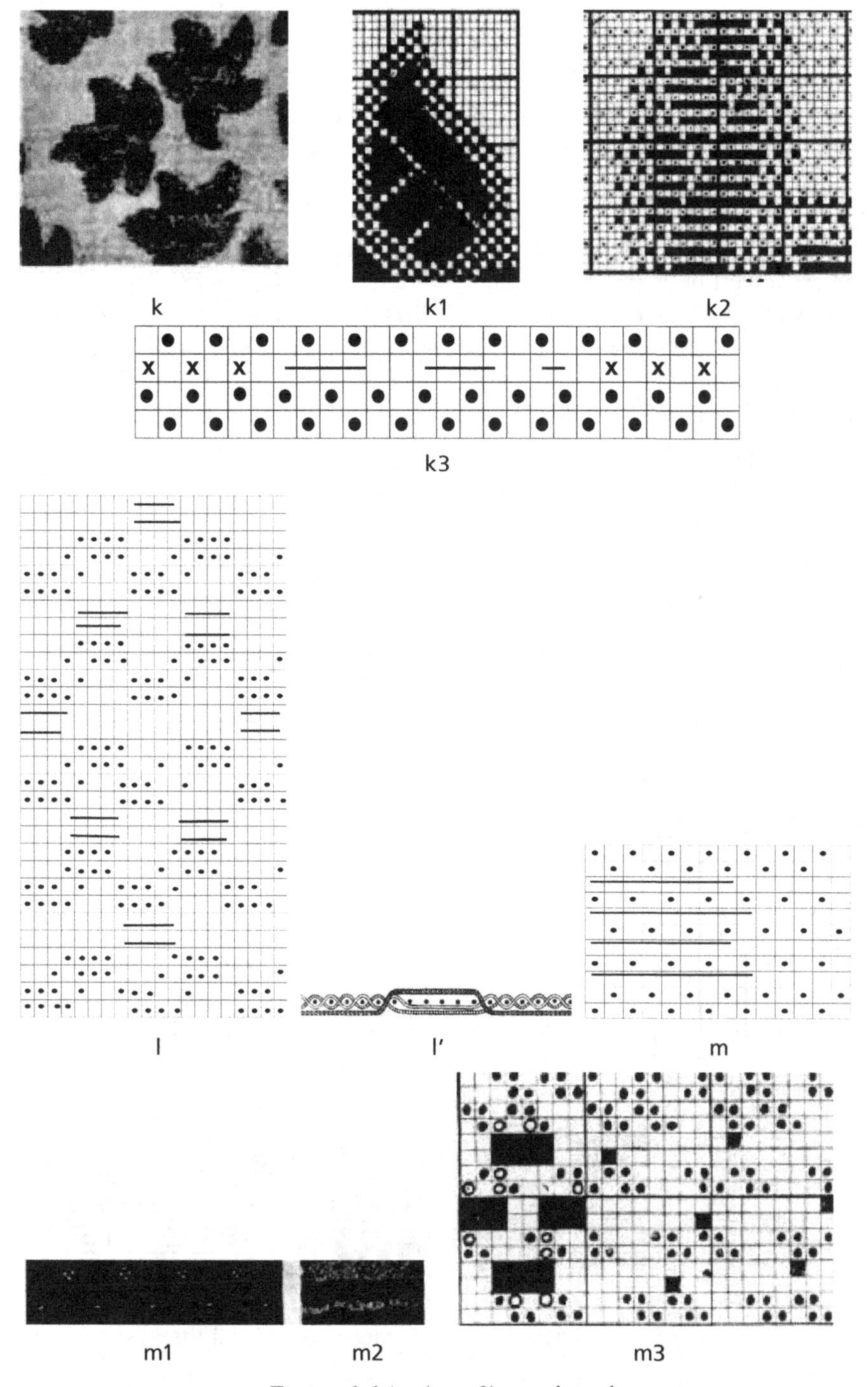

Figure 1.14 (a–m3) continued

or intermittent order. In the majority of the cases, we find disposal at regular intervals along the length of fabric and are woven on pit looms with special miniature shuttles controlled by weaver manually. The principle of cut-in is indicated in Figures 1.14k to k3, which show the designs in which the cross mark for extra weft indicates the plain form of binding with ground threads.

1.4.6 Ornamentation in extra weft figuring

Ornamental effects can be improved by selecting not only the RTP of ground and extra weft but also the type and nature of threads as extra weft. Greater solidity of extra weft can be obtained by discontinuing the ground weave picks at the place where extra weft floats, keeping the ground ends buried beneath the figure formed. Figure 1.14L′ and cross sections in Figure 1.14L′ show the situation of extra weft in which ground is modified hopsack and extra weft is disposed forming a motif.

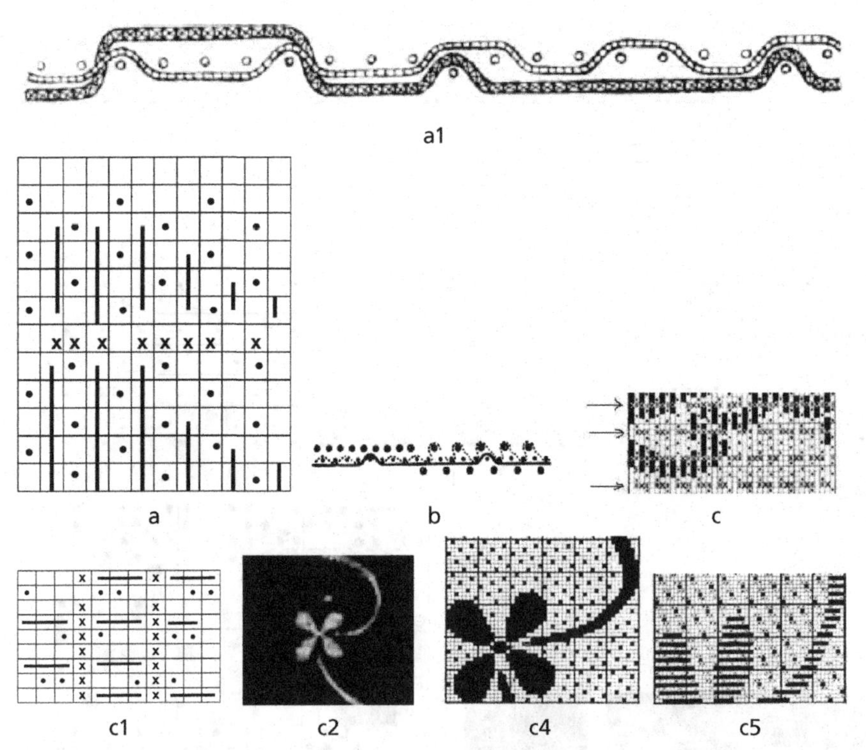

Figure 1.15 (a1–c5) Stitching by means of special threads

1.4.7 Stitching by means of special threads (ends/picks)

In extra thread figuring, it is found that sometimes the extra thread forms a long float and when simple means of binding become difficult, it is preferable to use certain extra threads in warp or weft direction depending on the case.

1.4.8 Stitching by means of special ends

When it is desired to retain the extra ends on the underside of a cloth without leaving them to float loosely between the figures, and when the ordinary method of stitching them between face floats of the ground ends is not feasible, the following system may be employed: special picks are introduced at intervals of 20 ordinary picks or if desired more frequently (Fig. 1.15a1 to c1). When they are introduced, the take-up motion must be rendered inoperative but unless the ordinary weft is very coarse they are not required to be different from the remainder of the weft. In most cases, they are same as the ordinary weft, and only perform a different function. It will be seen from the cross section of an extra warp fabric, through the warp, that on the special pick all the extra ends are raised, as are also the ground ends with the exception of those deliberately left down to provide a binding point for the special weft. In other forms of stitching, alternate extra ends may be raised on alternate stitching picks so that the first special pick binds in only the odd extra ends and the second only the even ones. The float of the stitching weft on the underside may also be made longer or shorter as required. The frequency of the stitching must not be too great; however, in such cases the dentations are liable to show in the face of the cloth. In Figures 1.15a to c, where the designs of Figure 1.15a and Figure 1.15c are the developed portions and Figure 1.15b is a cross section showing the use of special picks.

1.4.9 Stitching by means of special ends for extra weft

Similarly, in extra weft, special ends are used. Figure 1.15c1 shows a design where crosses indicate where the binding ends are left down, lines indicate the extra weft figuring, and dots indicate the ground weave in 2-up–2-down fashion in weft direction. The binding ends are down on all the figuring picks and are raised alternately on every nth ground pick.

1.4.10 Stitching in of extra weft fabrics

The principle of stitching of extra weft on the underside of the fabric in positions where it is not required for forming the figure is achieved by stitching with ground threads. In the case of those fabrics where the additional weight is not objectionable, the approach is followed. The method will give firmness

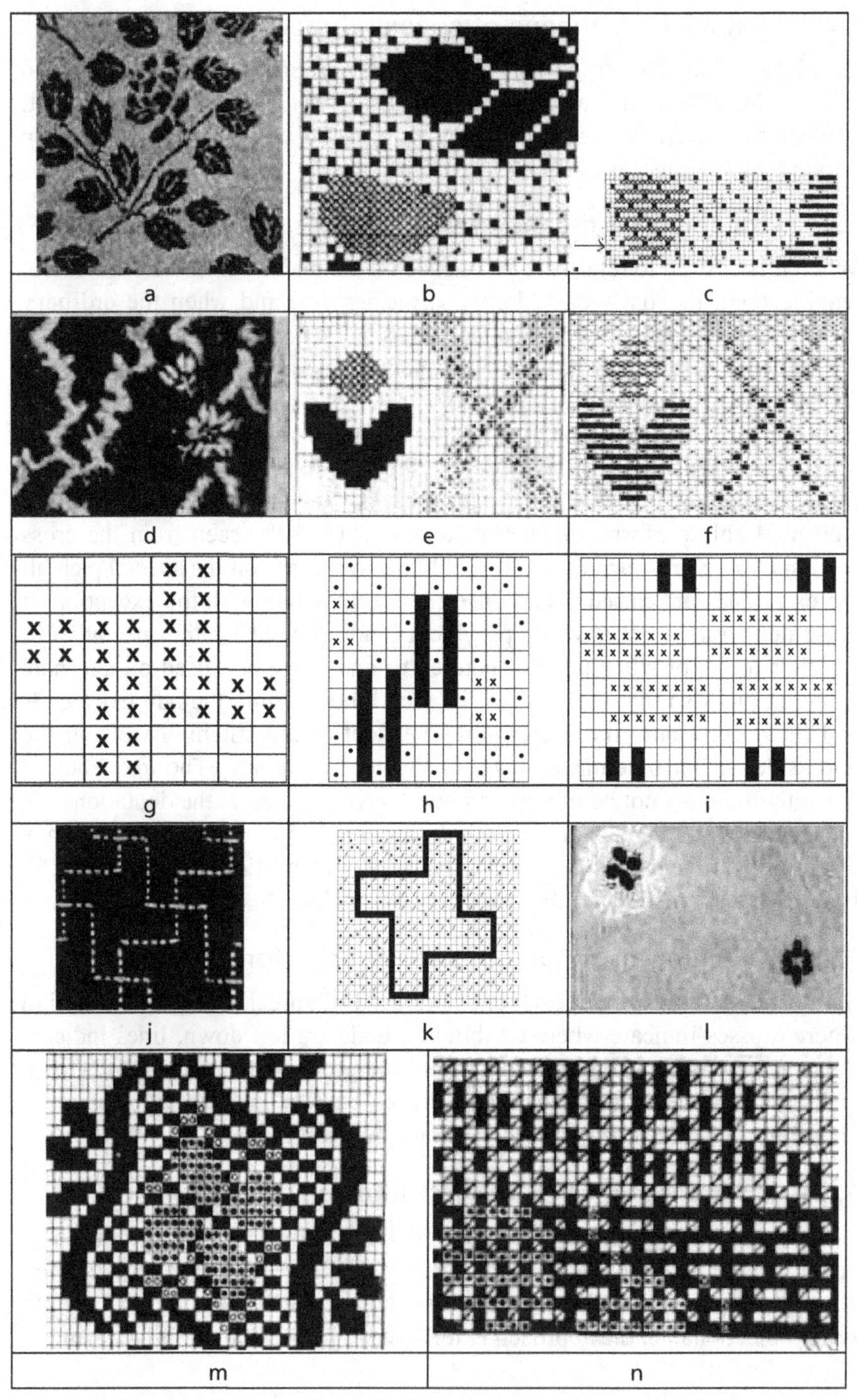

Figure 1.16 (a–n) Combined extra warp and extra weft effect

to the fabric and the ground weave may be plain or twill in 1:1 or 2:1 fashion. Figure 1.15c2 shows a sample of the fabric in which the extra weft is stitched and the design in Figure 1.15c4 is the developed portion and Figure 1.15c5 shows the design development of the design of Figure 1.15c2.

1.4.11 Combing ground float and extra thread float

In this class of fabric, the design is prepared in such a way that the ground float is combined with extra thread float and thus the binding of the fabric is achieved. Figure 1.16a shows the fabric and Figure 1.16b shows the portion developed and Figure 1.16c shows the weaves. The solid mark indicates the ground and cross indicates the extra thread figuring and the dot indicates the five-end sateen binding of the extra weft.

1.4.12 Continuous extra weft with two colours

In this class of fabric, two extra wefts are used in two colours. The final design will show the solid mark for first extra weft, crosses for second extra weft, and ground by dots. Figure 1.16 d shows the fabric, Figure 1.16e shows the development and Figure 1.16f shows the final design.

1.4.13 Combined extra warp and extra weft fabric

The combination of extra threads in both directions gives wide scope for ornamentation in an economical way when only one series of threads are involved in figure forming. Figures 1.16i to n show two examples of combined extra warp and extra weft. Such a fabric will need the loom equipment like dobby and box motions.

1.4.14 Chintzing

Ornamentation further in extra thread figuring is achieved by replacing one colour of extra thread by another in succeeding rows of design by a process called ‘chinzing’. This technique adds considerably to the variety of effect achieved without additional cost. Chintzing is often used in extra weft figuring where two distinct series of extra threads are inserted and two differently coloured extra weft figures can therefore exist side by side.

1.5 Differences between extra warp and extra weft

Extra warp figuring	Extra weft figuring
1. Consists of one series of weft and two or more than two series of warp known as ground and extra warps	1. Consists of one series of warp and two or more series of weft known as ground and extra wefts
2. Two or more warp beams may be required	2. Only one warp beam needed
3. No special box and take-up motions are required	3. Loom with box and special take-up motions are required.
4. Productivity of a loom is greater because only one series of picks is inserted so faster running loom can be used	4. Productivity is less because drop box looms have to be used
5. No theoretical limit to the number of colours that can be introduced	5. Maximum number of colours that can be introduced is dependent on which box motion is attached to the loom
6. With intermittent arrangement of extra ends either spotted or stripe patterns can be formed	6. Only spotted patterns are formed because stripe patterns lead to objectionable appearance of horizontal lines
7. Width of the repeat is dependent on the capacity of the jacquard	7. Width of the repeat is not dependent on capacity of the jacquard
8. In dobby weaving, drafts are usually complicated and number of healds are more	8. Drafts are simple and less number of healds required
9. Stronger yarn required for figure, and threads are not so soft, full, and lustrous	9. Weaker yarn can be used for figure; thus, threads are soft, full, and lustrous
10. Greater tension of extra ends in weaving, less contraction in length than width, extra warp effects show less prominently	10. Less tension on extra picks, therefore more contraction in width than length, the extra weft figuring most prominent
11. Removal of extra ends from the underside of the cloth is more difficult and costly	11. Removal of extra picks from underside is easy and economical

Extra warp figuring	Extra weft figuring
12. Stitching picks are sometimes used to stitch the extra ends floats on the back of the fabric	12. Stitching ends are sometimes used to stitch extra picks on the back of the fabric

1.6 Differences between backed clothes and extra thread figuring

Backed fabrics	Extra thread figuring
1. Extra series of threads known as 'backing threads' are used in order to increase the weight, strength of bulk, and warmth of the fabric	1. Extra series of threads known as 'extra threads' are introduced for forming figured effects on the surface of the ground texture
2. Backing threads are tied in such a way that they are not visible from face and they are only visible in the back side	2. Extra threads form prominent figured effects on the face side. In some cases, they are cut away from the back side, and in such case, they will be visible only from the face side
3. Here, there is no necessity for disposing the back threads	3. Extra threads must be disposed within portions where they are not required to form figures
4. Fabric is strong	4. Here, fabric is quite strong but extra threads fray out quickly
5. Wadding threads inserted to increase weight	5. No wadding threads inserted here
6. No need for special threads for stitching	6. Special stitching threads are required when the floats of the extra threads are too long
7. Backing threads are introduced in continuous order	7. Extra threads may be introduced in continuous or intermittent order
8. Used for bed sheets, rugs, and so forth	8. Used in manufacturing of dhoti and saree border extensively, and also for bed covers, and so forth

CHAPTER 2

Backed fabrics

2.1 Introduction

Backed clothes or fabrics as the name indicates will have some threads floating at the backside of the fabric and will not be seen to the observer. The backed principles of construction are employed for the purpose of increasing the strength, weight, bulk, and warmth-retaining qualities of a cloth and at the same time present a fine structure on the surface. A single cloth can be made heavy by using thick yarns and comparatively fewer threads per unit space, but this leads to a somewhat coarser appearance. By interweaving threads on the underside of a cloth, it is possible to obtain any desired weight combined with the fine surface appearance of a light single fabric. Backed fabrics are thus characterized by an additional series of either warp or weft solely inserted for increasing some of the cloth characteristics mentioned above. Backed fabrics can be termed as *semi-compound fabrics*. They occupy a position midway between 'simple' fabrics composed of one series each of warp and weft threads, and 'compound' fabrics, which contain two or more series each of warp and weft, as exemplified by all double clothes. Backed clothes are largely produced in worsted textures and are intended for boys' and men's clothing for which they are eminently adapted, as they are capable of yielding firm and compact though soft and warm textures. The principle of backing is shown in Figure 2.1. In the following diagram, the warp cross section is shown and the arrows show the binding for backed weft ends.

Backed fabrics are classified into:

1. Warp-backed clothes: Having two series of warp – face and back – and one series of weft.
2. Weft-backed clothes: Having one series of warp and two series of weft – face and back.

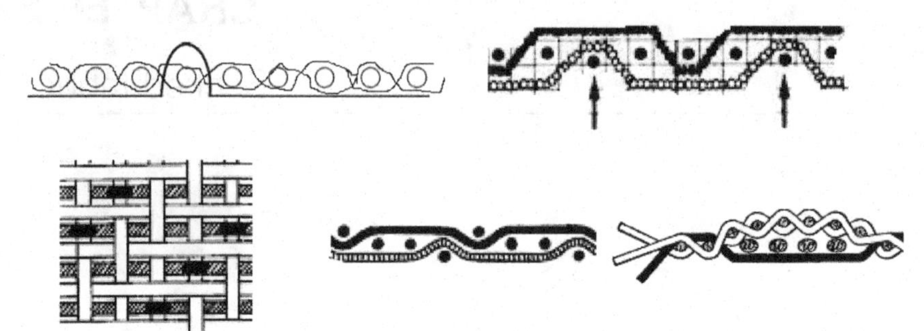

Figure 2.1 Principle of backed cloth formation

2.2 Warp-backed fabrics

These are the type of fabrics in which the backed thread will be positioned in warp direction as shown in Figure 2.2. *It should be noted that marks indicate warp up in all designs under this class.* In the following diagram, the weft cross section is shown and the arrows show the binding for backed warp ends.

2.2.1 Development of warp-backed cloth

The composition of the fabrics includes two series of warp, the 'face' warp and 'back' warp, with only one series of weft named 'face' weft. The standard orders of arranging the ends in warp-backed clothes are: 1 face:1 back, 2 face:1 back, and 3 face:1 back (there is no need to arrange the ends in even numbers); while in some cases, a backed weave is combined in stripe form with a single weave. For example, Figure 2.3 shows the simple case of warp-backed fabric with 3-up–1-down as face weave and 1 up and 3 down as back weave. Figures 2.3a, b show the face and back weaves, Figure 2.3c shows the superimposing and Figure 2.3d shows the development and Figure 2.3e shows the warp cross section. On the other hand, 'Figure 2.3f shows the draft, Figure 2.3g shows the double repeat of the weave and Figure 2.3h shows the weft cross section.

The construction of warp-backed fabrics (Fig. 2.3a to h) involves the following steps:

1. Selection of suitable face and back weaves
2. Arrangement of face threads and back threads as per the RTP
3. Insert face weave on face threads only
4. Insert back weave on back threads only
5. Tie up back warp to the face texture

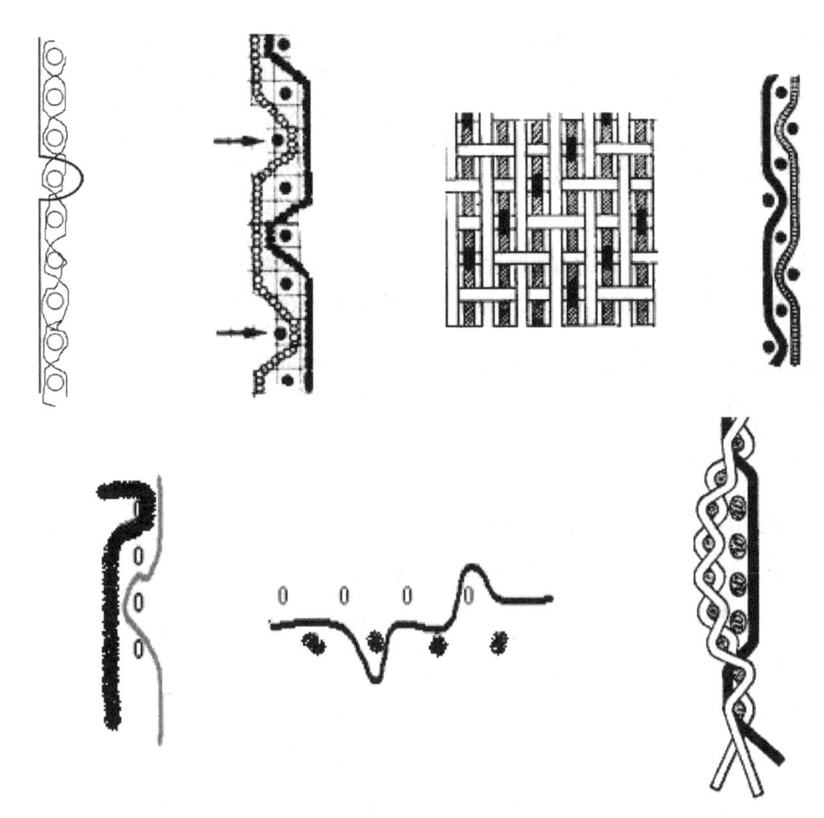

Figure 2.2 Principle of backing of warp threads in fabrics

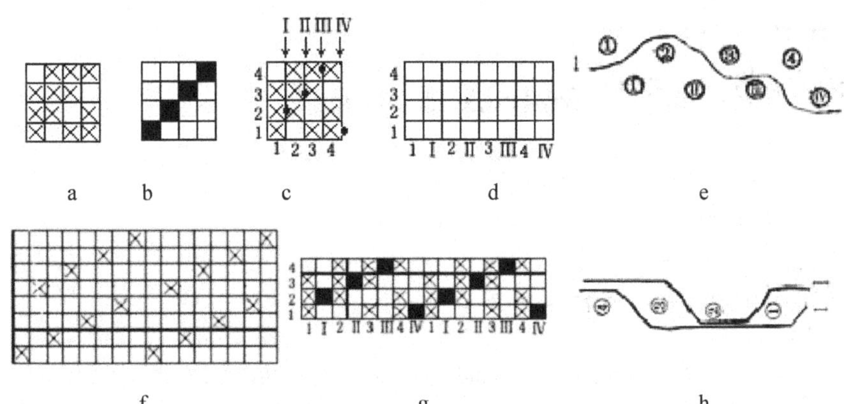

Figure 2.3 (a–h) Stages of construction of warp-backed fabric

The binding of 'back' warp threads must occur at such places as will ensure the binding points being properly covered by 'floats' of face warp threads. In other words, a 'back' warp thread should only be raised over a pick when the latter is passing beneath two or more 'floats' of face warp threads, otherwise the binding points will be liable to show on the face of the fabric.

2.3 How to find the repeat size of backed cloth designs – LCM (Least Common Multiple) approach

Let the face weave be 3/1 twill, back weave be 1/3 twill then, RTP (Relative thread of Proportion) will be 1:1 in warp as these are warp backed clothes (for illustration purpose)

Rf – Total warp ends in final repeat

R_y – Total weft ends in final repeat

R_m – Number of warp ends in face weave repeat

R_n – Number of warp ends in back weave repeat.

a, b – RTP of warp and weft.

R_y = LCM R_m & R_n

Rf = LCM of {LCM of R_m & a/a} and {LCM of R_n & b/b} (a+b)

$$Rf = \text{LCM of} \left\{ \left(\frac{4\,\&\,1}{1} \text{ and } \frac{4\,\&\,1}{1} \right) (1+1) \right\}$$

$$= 4 \times 2 = 8$$

R_y = LCM R_m & R_n = 4

2.4 Reversible warp-backed fabrics

2.4.1 Condition for reversibility

The face floats must be followed by back floats.

In reversible warp-backed clothes, we find that face warp floats are followed by back warp floats and the weaves selected will be perfectly reversible in nature.

Examples of reversible warp-backed fabrics are shown in graph designs.

Referring to the example Figure 2.4a, the details are as follows:

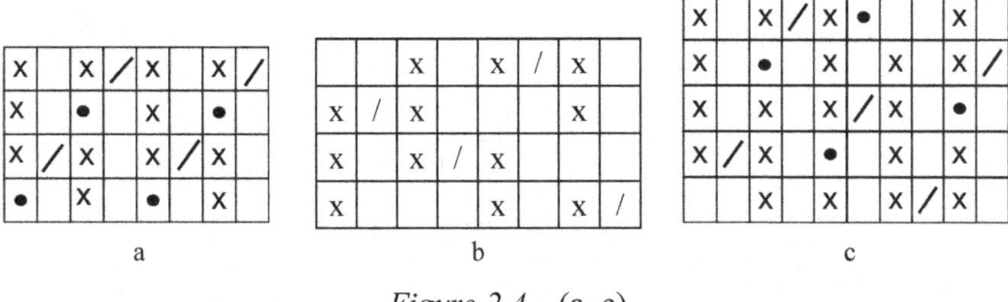

Figure 2.4 (a–c)

Figure 2.4a : Face weave: 4-end satinnette, back weave: 4-end weft sateen, RTP: 1:1 of F: B in warp direction. Total number of healds = 4 for face and 4 for back threads, loom equipment: dobby/jacquard.

Figure 2.4b: This is similar to the previous example but with different footing of face weave. Face weave: 4-end satinnette, back weave: 4-end weft sateen, RTP: 1:1 of F:B in warp direction. Total number of healds = 4 for face and 4 for back threads, loom equipment: dobby/jacquard.

Figure 2.4c: Face weave: 5-end sateen, back weave: 5-end weft sateen, RTP: 1:1 of F:B in warp direction. Total number of healds = for face and 4 for back threads, loom equipment: dobby/jacquard.

2.4.2 Backed fabrics with standard twill weaves

Backed clothes can be constructed with twill weaves as face and back weaves. In this class of fabrics, it is necessary to select all the face weaves in warp-faced condition only. Backed clothes are also constructed with face and back weave with different twills. Generally, face and back are selected in such a way that the RTP is 1:1 in warp. These are also found constructed with 2:1 RTP. However, the total number of picks in the final developed design will be the same as of basic face weave as backing will be warp direction only. Figure 2.4d shows 2-up–2-down as face weave, Figure 2.4e shows the development and Figure 2.4f shows 3 up and 1 down as face weave and 1-up–3-down as back weave. Figures 2.4 g and h show, respectively, the draft and peg plan for the weave selected in Figure 2.4d. Figure 2.4i shows the use of 4-end twill as face weave with two-warp-way repeats with 8-end sateen as back weave (5 move Number). Figure 2.4j shows the design with 2-up–1-down with two repeats as face weave. Figure 2.4k is to be completed by the reader. Figure 2.4l similarly shows the design with matty as face weave and the reader is directed to complete.

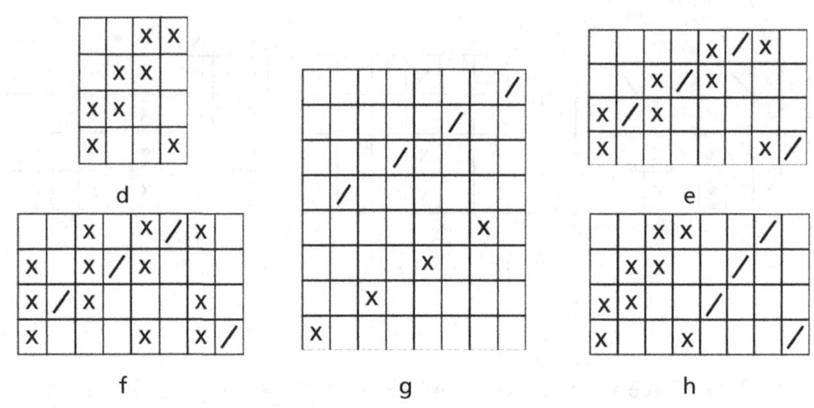

Figure 2.4 (d–h) Backed fabrics with standard twill weaves

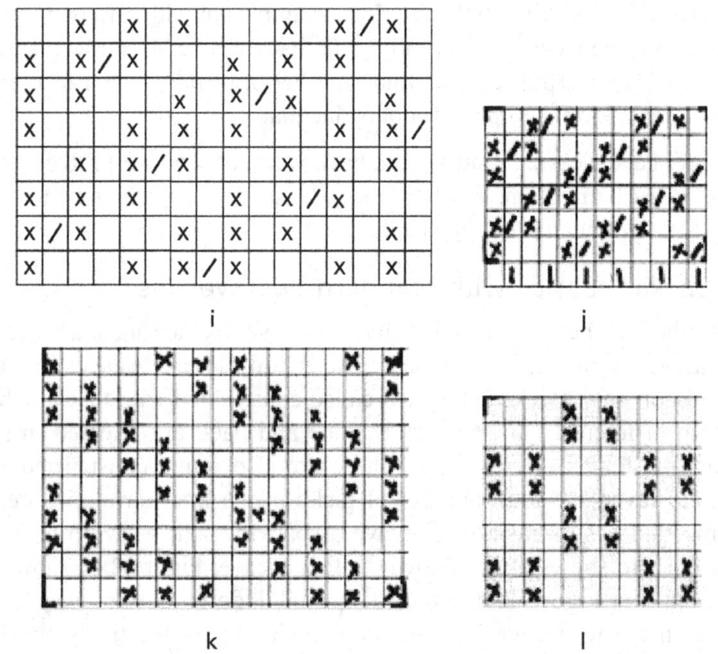

Figure 2.4 (i–l) Warp-backed clothes with standard twill weaves

2.4.3 Beaming and drafting of backed clothes

If face and back yarns are similar and the face and back weaves are of equal firmness, then a single beam can be employed. However, it is usual to employ separate beams for face and back because back threads usually interweave

			B
		F	
	B		
F			

Arranging the healds in alternate fashion.

			B
		B	
	F		
F			

Arranging the healds with all face healds at front and back healds at back fashion.

			F
		F	
	B		
B			

Arranging the healds with all back healds at front and face healds at back fashion.

Figure 2.5 Methods of drawing of backed clothes

with the weft than do face warp; thus, the back threads contract lesser than face warp (back warp threads float more) during weaving so they have to be contained on separate warp beams to facilitate independent regulation of tension. Also, back warp threads should pass from the warp beam to the healds at a lower level in a slightly lower plane than face threads, and should not be raised quite high as the latter during shedding. By observing these precautions unnecessary abrasive friction and chafing of warp threads will be avoided and any tendency of binding points to show on the face is thereby reduced (Fig. 2.5). *In drafting*, warp-backed designs simple patterns may be drawn straight over. In cases, however, where there is a difference in thickness or material between the face and backing ends, or if warp patterns are different for the two sides of the cloth, or if the face weave requires a special draft, it is better to draw each series through a separate set of healds. The back warp threads are preferably placed in the rear of the face healds, though the opposite may also be adopted. Figures 2.6a and b show the drafting and lifting plans.

2.4.4 Advantages of warp-backed fabrics

The advantages of warp-backed fabrics are:

1. Ordinary loom without any special devices can be employed.

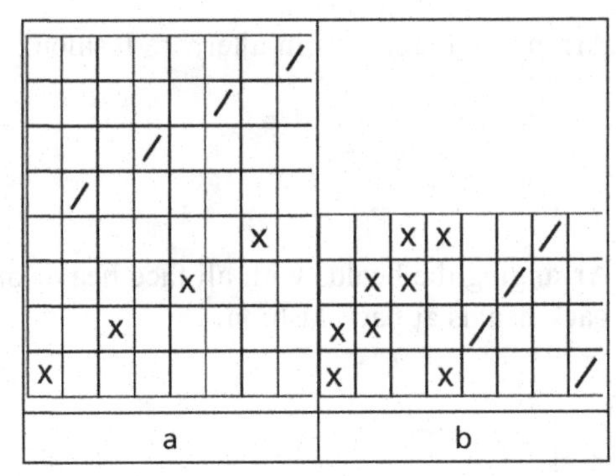

Figure 2.6 (a, b) Example for drafting of backed clothes along with peg plan

2. Reduction in weaving cost-less number of picks per centimetre (compared to weft-backed clothes) so increased production, lesser wages to weaver.
3. A more solid appearance can be given to the cloth by the formation of stripe patterns on the underside, which is impossible in weft-backed textures.
4. Due to the greater strength warp way, the clothes are superior from a structural point of view.

2.4.5 Disadvantages of warp-backed fabrics

The dis advantages of warp-backed fabrics are:

1. Two warp beams required.
2. On account of greater strain in weaving, however, such a low quality of backing yarn cannot be used as in weft-backing.
3. Drafts are complicated, and greater number of healds are required in producing similar effects.
4. Cost of drawing in is more because of greater number of ends.

2.5 Weft-backed fabrics

These fabrics are characterized by one series of warp and two series of weft – face and back. The standard orders of arranging the picks in theweft-backed clothes are: 1 face:l back, 2 face:l back, 3 face:l back, 2 face:l back, 2 face:2 back, and 4 face:2 back. The last two arrangements are used in place of the first two when a different kind of backing weft from face weft has to be used in looms with changing boxes on one side only. 2:1 and 2:1 arrangements are the ones most commonly used. In 1:1 clothes, usually both face and back are of similar thickness and used in the production of fine fabrics. In 2:1, there are half as many back picks as the face so the backing yarn is usually of cheaper and inferior quality when compared to face picks. Thus, the cloth produced with 2:1 arrangement is less attractive because the underside appears coarser. *It should be noted here that the marks in the design indicate the weft up.*

The construction of weft-backed fabrics is governed by the same principles as those which govern the construction of weft-backed fabrics as regard the method of securing the extra series of threads to the face texture. The design should be such that each warp thread passes beneath not less than two contiguous picks of weft (but with each warp thread passes under different picks) at least once in each repeat of the design.

2.5.1 Reversible weft-backed clothes

If face and back weaves are reversible in nature, the backed clothes produced are known as reversible weft-backed clothes. Generally, these are constructed by selecting reversible weaves in 1:1 or 1:2 or 2:1 RTP. Sometimes, the use or 3:1 is also recommended. Depending on the RTP, the denting plan in warp-backed and box motion pattern chain is designed. Examples shown in Figure 2.7.

2.5.2 Construction of weft-backed clothes with twill weaves

Like warp-backed clothes, weft-backed clothes are also constructed with twill weaves in different RTP, and accordingly, the repeat size varies. In weft-backed clothes, picks are more depending on the RTP and number of ends will be equal to the basic weave used as face weave. Figure 2.8a shows 2/2, 2/2 Z twill as face and back weave 1/3,1/3 Z twill. Here, it may be necessary to have only one stitching or binding point but Figure 2.8i shows two binding points for each back pick. The purpose is to increase firmness by selecting the binding at two points instead of one point. As per the principle, only one point per pick is sufficient. Figure 2.8b show 3-up–1-down face weave, Figure 2.8c shows 2-up–2-down face weave arranged with 2:2 RTP, Figure2.8d shows 5-end twill as face with two repeats in weft, Figure 2.8e shows 2-up–1-down and 1-up–1-down as face and the reader is directed to complete the weave. Figure 2.8f show 2-up–2-down matty as face and the reader is directed to modify the design so that the second and fourth back weft will be in contact.

In Figure 2.8g shows 2-up–1-down and 1-up–1-down with two repeats as face; Figure 2.8h shows 5-end twill as face with one repeat, Figure 2.8i shows 4-end twill as face with 2:2 RTP with single binding point, whereas Figure 2.8j shows the similar arrangement but with two binding points for back weave, Figure 2.8k is similar but the RTP is 3:1, Figure 2.8l shows 4-end

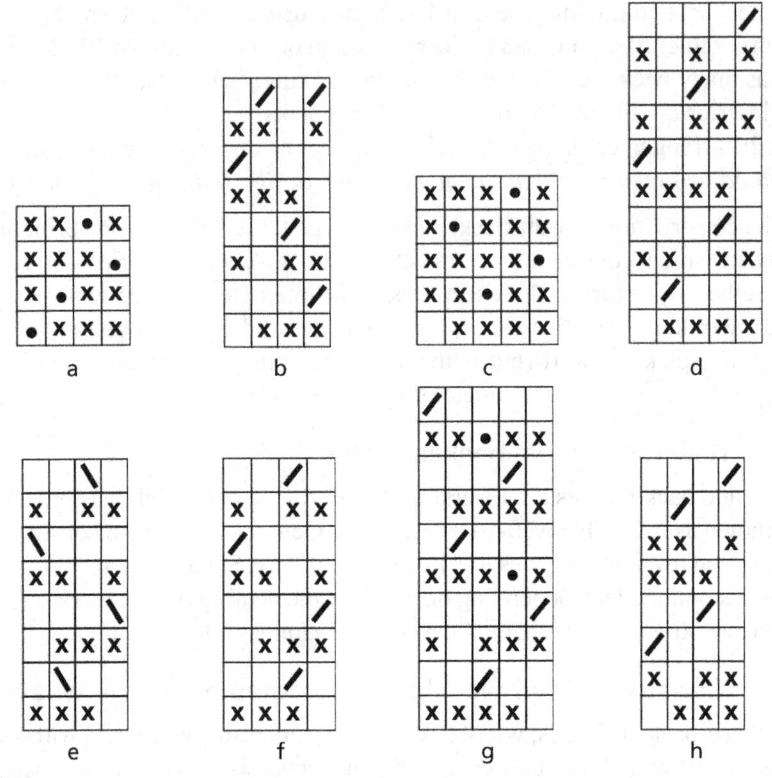

Figure 2.7 (a) Face weave: 4-end satin or satinnette, (b) is the Development of 2.2 a with back weave. 4-end weft sateen, (c) Face weave 5-end satin, back weave 5-end sateen, (d) is the Development of Figure 2.7c with back weave, (e) is Similar to Figure 2.7a but with different footing for face weave, (f) is Similar to Figure 2.7a but with different footing for back weave, (g) Face weave as 5-end satin and back weave as 5-end weft sateen, (h) Face weave 4-end satinnette, back weave 4-end sateen but with 2:1 RTP for face:back. (Note: When back pick is inserted, all warp threads are raised excepting those which are required to pass under it for the purpose of binding it to the fabric. For best results, all warp threads should, if possible, be utilized for binding the back picks to the face texture; the binding points should be uniformly distributed).

twill with two repeats, Figure 2.8m shows 6-end twill as face weave, Figure 2.8n shows 4-end twill as face weave but back weave is selected on 8-end sateen with 5 as move number, Figure 2.8o is also with 4-end twill but sateen

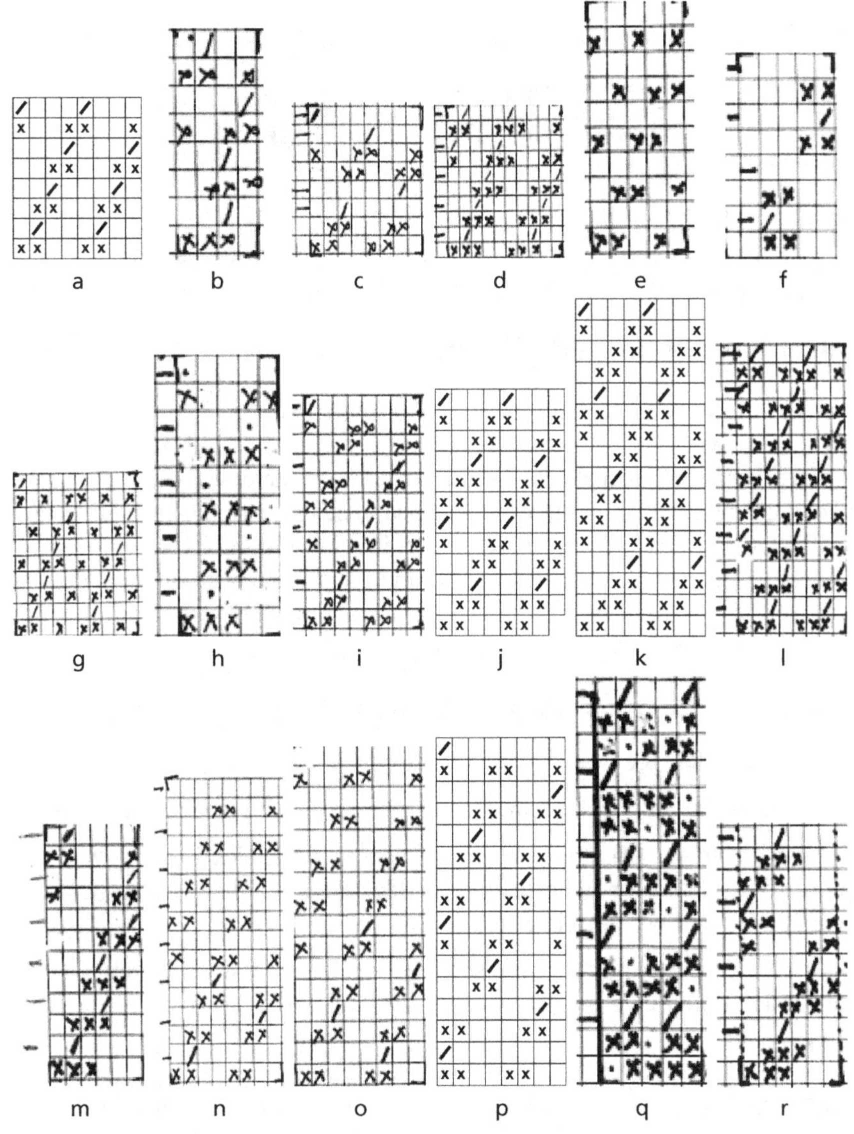

Figure 2.8 (a–r) Weft-backed clothes with twill weaves

as back weave with 3 as move number, Figure 2.8p as a design with matty as face, Figure 2.8q is with sateen as face but with 2:1 RTP, and Figure 2.8r as 6-end twill as face with 2:1. The reader is directed to draw all the weavers' plans for these designs.

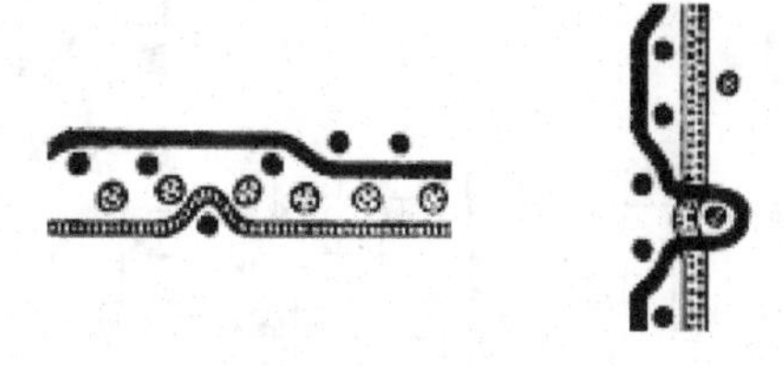

Figure 2.9 (a, b) Principle of wadding in backed fabrics

2.6 Wadded backed clothes

The role of wadding thread remains the same irrespective of the structure considered. Wadding is the threads added to the basic design with a view to improve the warmth nature economically and also to increase the fabric weight. Wadding is also aimed to improve the appearance of the clothes. Wadding can be affected in warp and weft directions. If it is introduced in warp direction, we need a separate heald shaft to guide and control the position of wadding. However, if it is used in weft direction, a separate shuttle is required with patterning mechanism.,

2.6.1 Constructional steps of wadded backed clothes

1. Select the face weave and back weave
2. Mark the RTP and mount respective weaves on respective threads
3. Indicate the lifts for wadding end on all back pick in warp wadded weft-backed clothes or indicate lifts for face ends on all wadding picks in weft wadded and warp-backed clothes.
4. Draw the wadding ends in a separate heald in warp wadded weft-backed cloth and mention note the use of separate shuttles for back pick.
5. Draw face ends in face heald and back ends in back healds.

2.6.2 Method of arranging wadding at the centre of two layers

Generally, it is well known that wadding threads are not seen to the observer and in rare cases like piques, wadding or backing threads are connected to base or body. If it is in warp direction, lift all the wadding ends on back pick and lower them on face pick (Fig. 2.9) so as to make them to lie at centre.

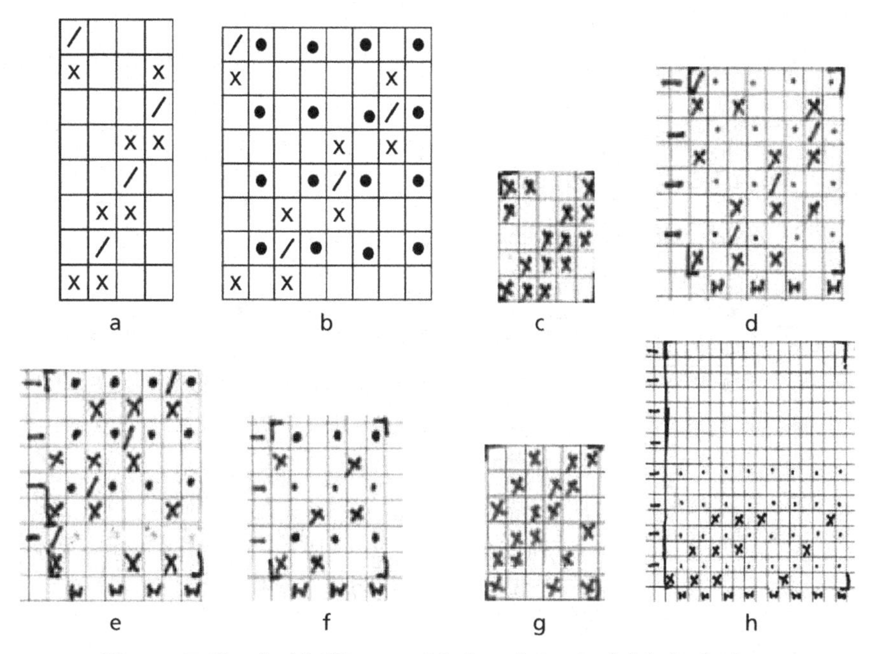

Figure 2.10 (a–h) Warp wadded weft-backed fabric design

If wadding used is a weft, lift all face ends on wadding pick so as to make wadding pick to lie at centre.

How many Number of wadding picks or ends to be used?

Normally, the number of wadding ends required will be equal to number of face ends and number of wadding picks will be equal to number of face weave picks.

2.7 Weft-backed warp wadded clothes

Choosing 2/2 Z twill as face and 1/3 Z twill as back weave with four wadding ends, weft-backed warp wadded clothes are developed as shown in Figures 2.10a and b, respectively. Note the lifts for wadding ends on back pick. Figure 2.10c shows a basic weave of 5-end twill and Figure 2.10d shows the development of Figure 2.21c. Figure 2.10e shows the developed design with 3/1 twill as face weave and Figure 2.10f is also a 4-end twill with 2/2. Reader is advised to develop the design for Figure 2.10g. The next design is shown in Figure 2.10h with 3/2,1/2 and the design is to be completed by reader.

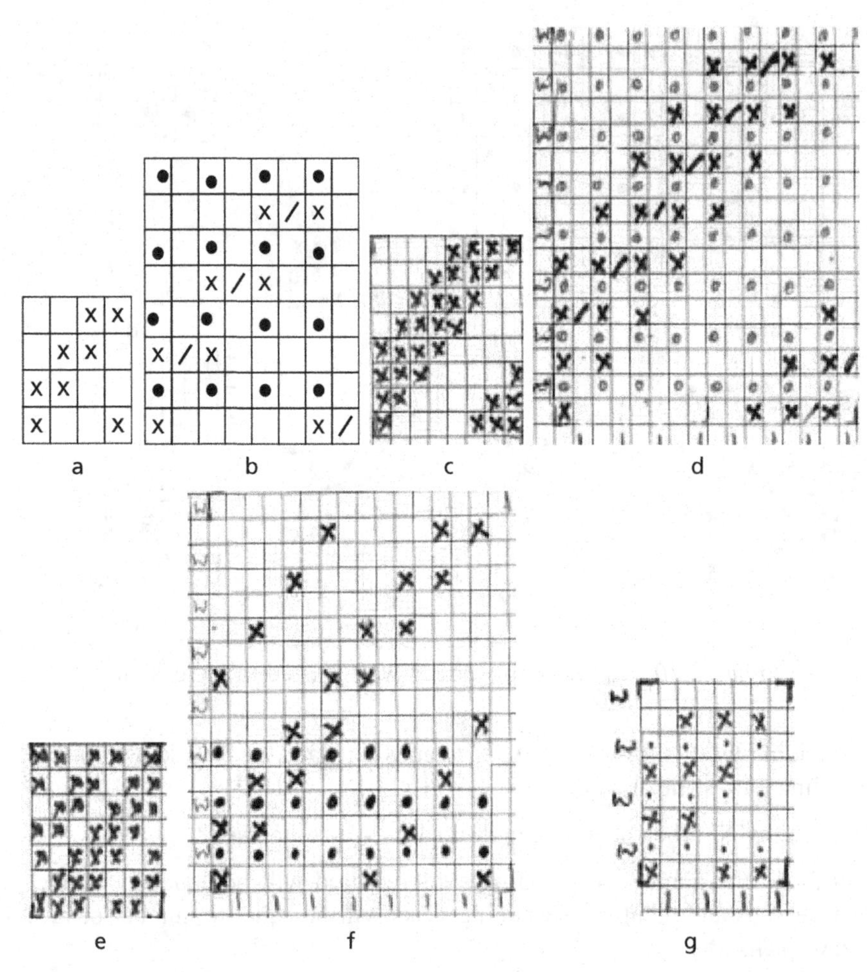

Figure 2.11 (a–g) Warp-backed and weft wadded fabric design

2.8 Warp-backed weft wadded cloth

Figures 2.11a and b show the arrangement in which four wadding picks are represented in 1:1 RTP in warp and weft directions. Lift all the face ends on all wadding picks. Figure 2.11c shows 4/4 weave and Figure 2.11d shows the development of Figure 2.11c. The reader has to develop the design for Figure 2.11e. The design of Figure 2.11f shows 2/2,1/3 face weave and the reader is directed to complete the design. Using face weave 3/1, the backed cloth is developed as shown in Figure 2.11g.

2.9 Imitation effects in backed clothes

It is possible to use colours in warp and weft for face and back threads and imitation effects can be produced in which the face and back will have similar floats or face and back floats to make a reversible fabric. The imitation effects are produced in warp or weft directions. Figure 2.12 shows both weft-backed and warp-backed clothes cross section in which the face and back threads will interlace equally. Any weave can be taken (should be square in shape) and can be imitated on one less or one more than the twice the number of threads in the basic weave. In Figure 2.12, weft imitated backed cloth for a weave 2/2 Z twill on 7 × 7 (Fig. 2.12a) and the same weave is imitated on 9 × 9 (Fig. 2.12b) and Figure 2.12c shows 5 × 5 imitation of 2/2 twill. Warp imitation is shown in Figure 2.12d on 7 × 7. Similarly Figure 2.12e shows the imitation on five ends and picks.

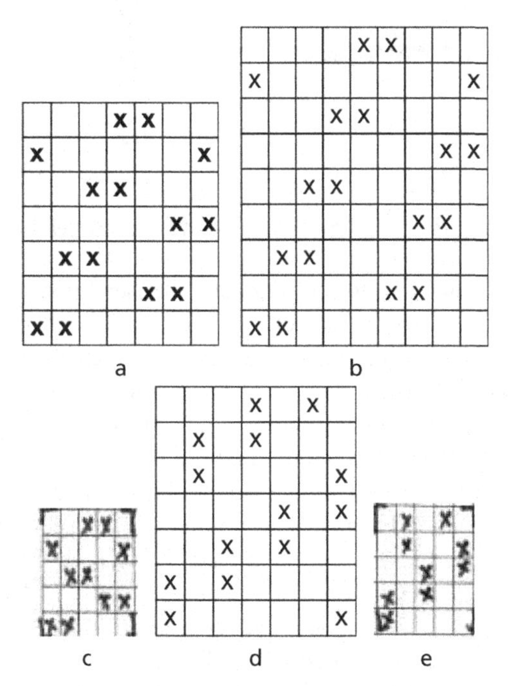

Figure 2.12 (a–e) Imitation-backed fabrics: Weft imitation and warp imitation

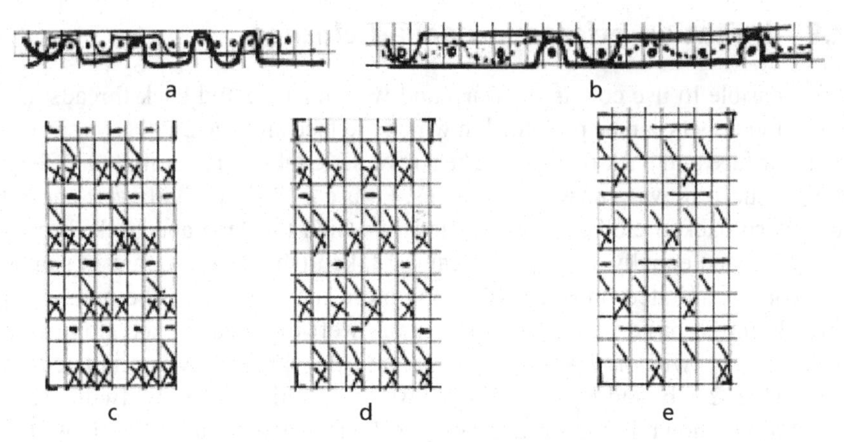

Figure 2.13 (a–e) Interchanging backed clothes

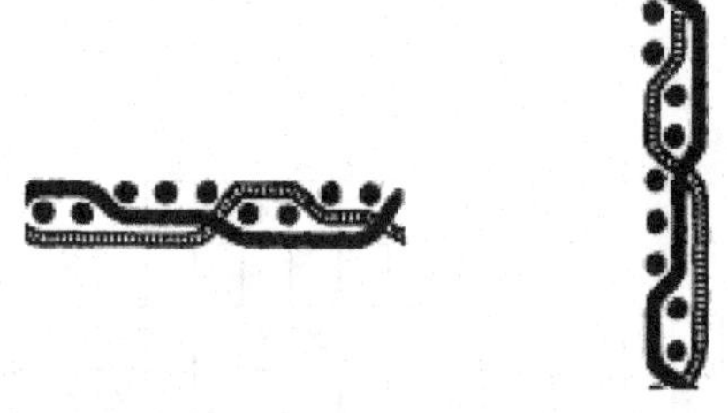

2.10 Interchanging effects in weft-backed fabrics

These clothes are chiefly used for blankets, dresses gowns, and rugs. The weave is same in every part of the cloth, and a weft surface is produced on both sides. The design is due to the manner in which differently coloured wefts are interchanged from one side to the other, a dark figure on a light ground on one side corresponding with a light figure on a dark ground on the other side. Generally, the wefts should be brought about more equally to the surface on both sides in order that one side will not appear different than the other, this being particularly the case when the cloth is seen on both sides at the same time. A raised finish is applied alike to both the face and back, and when woollen weft is used, the shrinkage in width ranges from 15 to 30%. The warp is invariably cotton. The felted and raised finishes cause the cotton ends to be entirely concealed and give a soft feel to the cloth. Cheap clothes are made entirely of cotton, the flannelette kind of weft being used. The

weaves for the figure and ground are constructed upon the same principles as weft-backed reversible designs (Fig. 2.13).

2.11 Differences between backed clothes and extra thread figuring

Backed fabrics	Extra thread figuring
1. Extra series of threads known as 'backing threads' are used in order to increase the weight, strength of bulk, and warmth of the fabric.	1. Extra series of threads known as 'extra threads' are introduced for forming figured effects on the surface of the ground texture.
2. Backing threads are tied in such a way that they are not visible from face and they are only visible in the back side.	2. Extra threads form prominent figured effects on the face side. In some cases, they are cut away from the back side, in such case, they will be visible only from the face side.
3. Here, there is no necessity for disposing the back threads.	3. Extra threads must be disposed within portions where they are not required to form figures.
4. Fabric is strong.	4. Here also fabric is quite strong but extra threads fray out quickly.
5. Wadding threads inserted to increase weight.	5. No wadding threads inserted here.
6. No need for special threads for stitching.	6. Special stitching threads are required when the floats of the extra threads are too long.
7. Backing threads are introduced in continuous order.	7. Extra threads may be introduced in continuous or intermittent order.
8. Used for bed sheets, rugs, and so forth.	8. Used in manufacturing of dhoti and saree border extensively, also for bed covers, and so on.

2.12 Differences between warp-backed and weft-backed clothes

Warp backed	Weft backed
These fabrics are less soft and loft to handle when compared to weft backed.	Softer and more lofty in handling This may be due to weft contains less twist and being under less tension than the warp.
As two series of warp threads exist, it requires two warp beams and no drop box.	As two series of threads exit in weft it requires one warp beam and drop box.
Warp-backed fabric can be economically produced (cheaper) due to less picks per centimetre	It is costlier as it is necessary to insert more picks per centimetre
A more solid appearance can be given to the cloth by the formation of stripe patterns on the underside.	Impossible to produce a solid appearance.
Due to two series of threads in warp, the warp way strength will be high.	Higher weft way strength due to two series of threads in weft.
These fabrics are very strong from a structural point of view.	These are weak and inferior from a structural point of view.
Fabric demands good and high quality of yarn as it cannot withstand greater strain in weaving.	Low quality of backing yarn can be used in weft due to less strain on yarn.
Drawing operation may be high cost as it involves more RTP in warp.	Drawing in is cheaper due to less number of ends.
Drafts are usually more complicated, and a greater number of healds are required in producing similar effects.	Drafts are simpler.
The standard order of arranging the ends in warp backed clothes are 1 face to one back, 2 face to 1 back, and 3 face to 1 back.	The standard orders of arranging the picks are 1 face to 1 back, 2 face to 1 back, 3 face to 1 back, 2 face to 2 back, and 4 face to 2 back.

CHAPTER 3

Double-layered structures

3.1 Introduction

Double clothes are those advanced fabrics with at least two series of warp and weft threads each of which is engaged primarily in producing its own layer of cloth, thus forming a separate face cloth and a separate back cloth. Thus, double clothes may be included under the heading of compound fabrics. The two or more clothes may be of the same or different weaves; they may be separate from each other, bound only at one selvedge to open out into one double width cloth; or they may be bound at both selvedges to give a bag or tube effect; or they may be bound together all over the fabrics to form one solid cloth, with the same or different weaves, for the face cloth and the back cloth. The purpose of weaving together two clothes may be entirely utilitarian, such as the improvement of the thermal insulation value of a fabric in which a fine, smart face appearance is necessary; or it may be aesthetic in intention for which purpose the existence of two series of threads in each direction improves the capacity for producing intricate effects dependent upon either colour or structural changes.

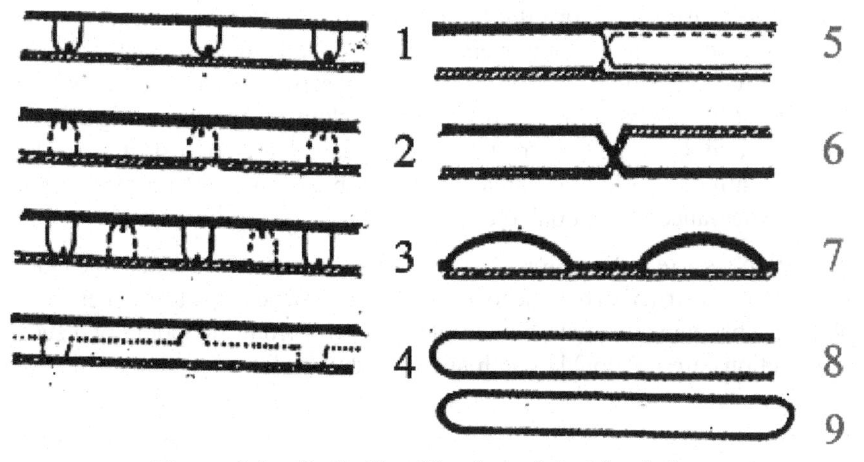

Figure 3.1 (1–9) Classification of double clothes

3.2 Classification of double clothes

3.2.1 Self-stitched double clothes

Fabric contains two series of both warp and weft, and stitching of face layer to back layer is accomplished by occasionally dropping a face end under a back pick, or by lifting a back and over a face pick, or by utilizing both systems in different portions of the fabric, which is illustrated below:

1. **Centre-stitched double clothes:** In these fabrics, a third series of threads is introduced either in the warp or in the weft direction whose entire function is to stitch the two otherwise separate layers of cloth together. The centre ends lie between the face and back cloth and for the purpose of stitching oscillate at regular intervals between the face and the back pick, thus achieving the required interlayer cohesion as shown below.
2. **Double cloth stitched by thread interchange:** These structures are similar to the first category in as much as they do not contain an additional series of stitching threads. However, they are distinguished from the self-stitched fabrics by the fact that the stitching of the face and the back cloth is achieved by frequent and continuous interchange of same thread elements between the two cloth layers. Thus, on some portions of the cloth, the face ends may be made to interweave with the back picks and the back ends with the face picks as shown below.
3. **Double clothes stitched by cloth interchange:** In this class of constructions, the principle of the interchange is taken one step further them in the above category and complete cloth layers are made to change places as shown in the figure. As stitching between two fabrics occurs only at the point of cloth interchange, the degree of cohesion in this type of cloth depends on the frequency of interchange.
4. **Alternate single and double ply construction:** In some fabrics, the constituent thread components are occasionally merged together into a heavily set single cloth and occasionally separated into distinct layers to form figures areas of open double cloth on the firm single cloth ground. Usually, the effect depends upon a degree of distortion as the crammed single cloth areas tend to separate out, thus affecting the appearance of the double cloth 'pockets'.

In addition, as already mentioned, some clothes are produced on double cloth principle of construction but due to the deliberate absence of stitching between, the layers become single cloth upon their removal from the loom. Two such constructions are the double width and the tubular cloth.

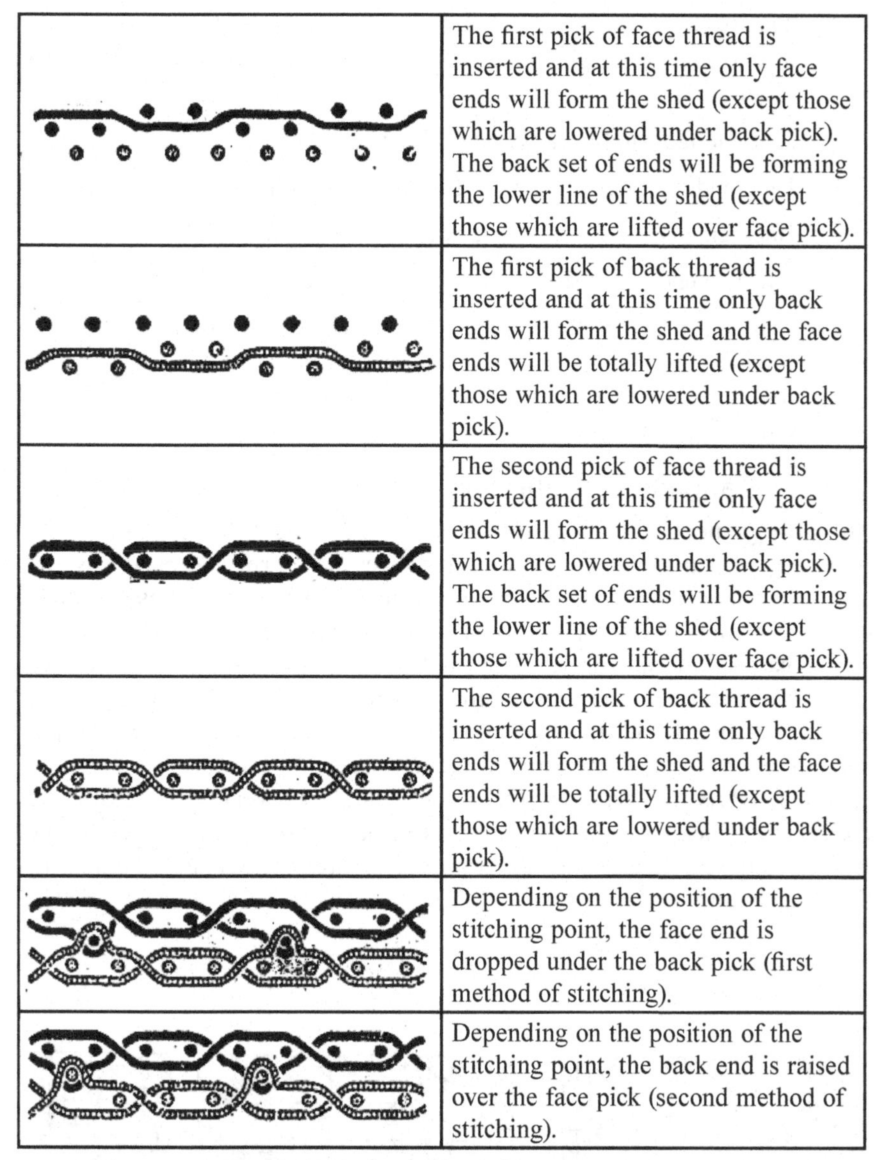

	The first pick of face thread is inserted and at this time only face ends will form the shed (except those which are lowered under back pick). The back set of ends will be forming the lower line of the shed (except those which are lifted over face pick).
	The first pick of back thread is inserted and at this time only back ends will form the shed and the face ends will be totally lifted (except those which are lowered under back pick).
	The second pick of face thread is inserted and at this time only face ends will form the shed (except those which are lowered under back pick). The back set of ends will be forming the lower line of the shed (except those which are lifted over face pick).
	The second pick of back thread is inserted and at this time only back ends will form the shed and the face ends will be totally lifted (except those which are lowered under back pick).
	Depending on the position of the stitching point, the face end is dropped under the back pick (first method of stitching).
	Depending on the position of the stitching point, the back end is raised over the face pick (second method of stitching).

Figure 3.2 Formation of Double Cloth

When the structure of double clothes is being built up, the following rules must be strictly followed out to give the required results:

1. Face ends and face picks must be arranged in definite order with back ends and back picks.

2. The two series of ends should be drawn through healds in such a way that one series may be operated quite independently of the other series.
3. Suitable face and back weaves are to be selected.
4. Face ends should weave with face picks according to the face weave.
5. Back ends should interweave with picks according to back weave.
6. All face ends are lifted when back picks are put in.
7. If the two clothes formed are to be stitched, lift up a back end into the face cloth on a face pick or drop a face end over a back pick, such that the stitching point is between two 2-face warp floats and thus get covered when woven.

3.2.2 Arrangement of threads

The arrangement of threads in the cloth is dependent on the end result required. The arrangement of weft is however determined partly by the type of loom used to weave the cloth. The most common varieties of double cloth are arranged in warp and weft, 1 face, 1 back, and 2F:1B. For looms with boxes on one side only, and when the back weft is different from face weft, similar effects may be obtained in many weaves by changing the wefting to 2F:2B and 4F:2B. Clothes which require a very fine face are sometimes arranged 3F:1B, mixed order arrangement is also possible, for example, 1F:1B in warp and 2F:1B in weft and vice versa. In deciding on the relative thicknesses of the face and back yarns, a good rule to follow is to have the relative counts about proportionate to the relative number of the threads/unit space.

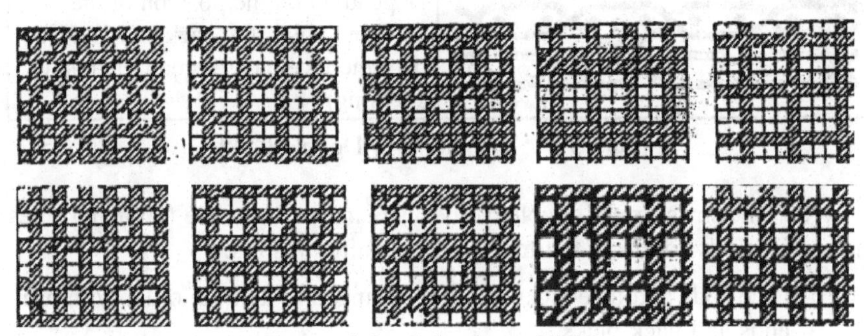

Figure 3.3 Selection of RTP

3.2.3 Selection of weaves: In general

1. If threads are arranged in equal proportions, the back weave is usually same as face weave or at least has the same number of relative intersections (e.g. 2/2 and 1/3).
2. In other cases, as a rule, back weave has more intersections than face weave in order to compensate for reduced number of threads.

   For example: 2F:1B, face weave = 2/2, back weave, plain.
3. Most regular effect is obtained by having the repeats of the face and back weaves equal, or one a multiple of the other.

   For example, 1/3 twill is unreliable for backing the 2/3 twill unless the threads are arranged irregularly in the proportion of 5 face-to-4 back.

Stitching: Stitching points should be properly placed such that they have no effect on the appearance on either the face or the back cloth.

1. **Raising back end on face pick:** This should be done only when the back end is away from the underside of the back cloth and the pick over which the tie is made must be away from the face of the top cloth.
2. **Dropping face end on back pick:** Both the face end and the back pick must be away from their respective surfaces.
3. **Face weave:** Warp satin or warp-faced twill, then lift back end on face pick.
4. **Face weave:** Weft sateen or weft-faced twill, then drop face end under back pick.
5. When there is a choice between the two methods, lifting back end on face pick should be preferred.

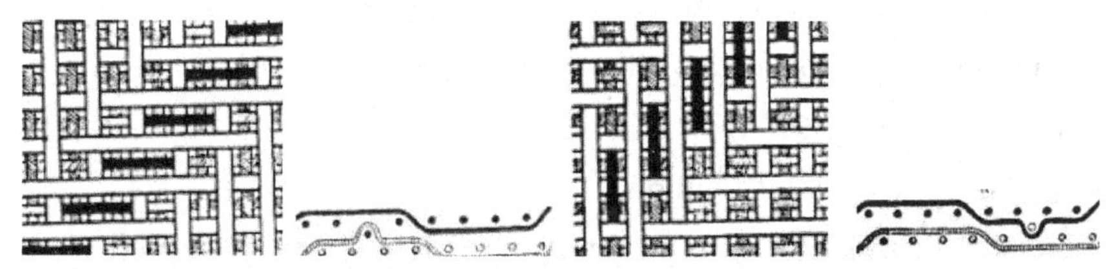

Figure 3.4 Principles of tying in double cloth

3.2.4 Selection of RTP in double clothes

Selection of RTP depends on the size of face and back weaves. When special threads like wadding or stitching are added, the RTP also vary and following table gives us an idea about the selection of RTP.

S No.	Face weave	Back weave	Structure	RTP in warp	RTP in weft
1	2/2 Z twill	6-end sateen	-	FBFBBFBBFB	FBFBBFBBFB
2	2/1, 2/3 stitched hopsack	2/2 S twill on 8 × 8	Warp wadded	FWB,1:1:1	FB, 1:1
3	2/2 skip twill	2/2 S twill	–	FBFFBFFBFFBF	FB, 1:1
4	2/3 matty	2/2,2/2 S twill	Weft wadded	FBBFBBFBBFBBF	FWBWFWBB WFWBB FWBBWFWB
5	3/3 Z twill	2/2 Z twill	Centre warp stitched	FBFFB*S*FFBFB	FBFFBFFBFB
6	3/1 matt	3/3 S twill	Centre weft stitched	FBBFBBFBBF	FBBFB*S*BFBBF

S - Stitching thread

3.3 Development of simple double clothes (self-stitched)

In this class of double clothes, the layers are connected to each other by self-concept without employing a separate stitching or binding thread. This is achieved either by dropping certain face ends under back picks or raising certain back ends over face picks. The systematic way of construction of self-stitched double clothes includes the following steps.

Example 1. Figures 3.5a1 to k4 illustrate the steps listed:

a) Face weave 3/3 Z twill on 6 × 6.

b) Back weave 3/3 Z twill on 6 × 6.

c) The final repeat size based on RTP 1:1 in warp and 1:1 in weft.

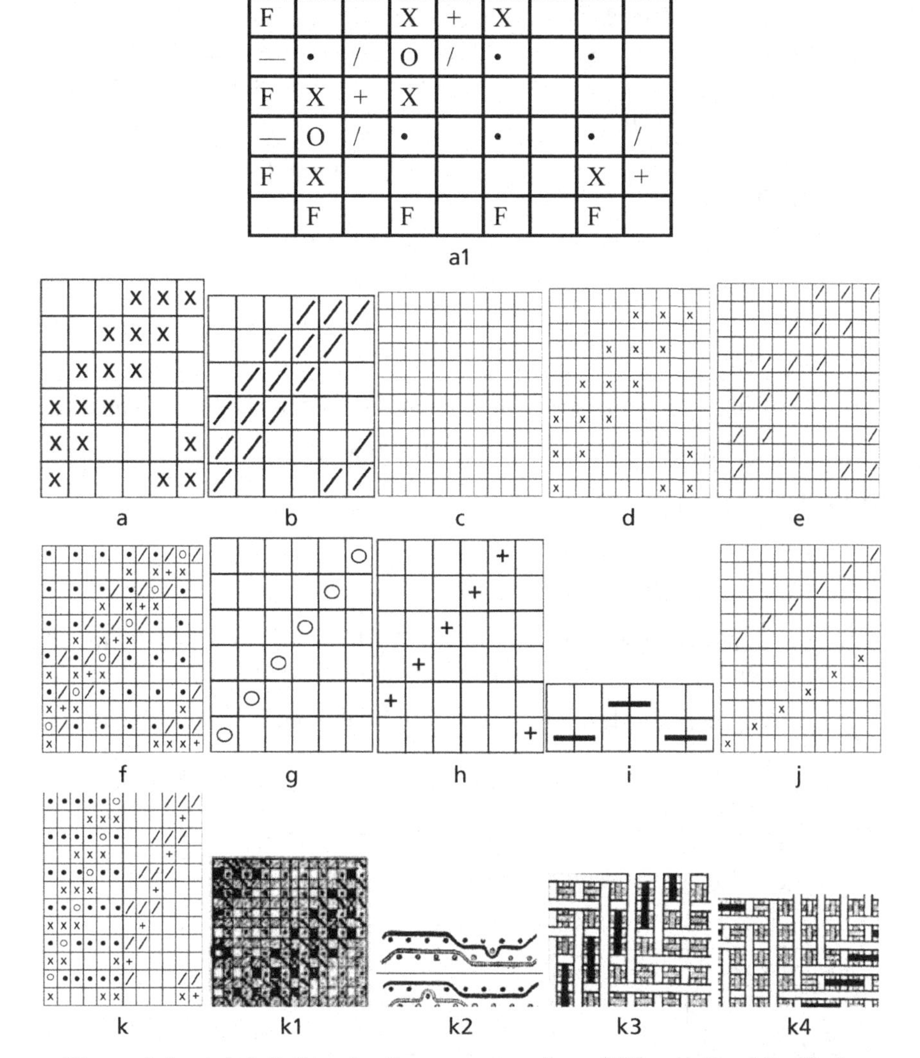

Figure 3.5 (a1–k4) Step by Step construction of Simple Double Cloths

d) Transfer of face weave on face ends and picks.

e) Transfer of back weave on back ends and picks.

f) Developed design with proper lifts for all face ends on back picks except at the places of stitching for the formation of top layer of shed and stitching points wherein face is connected to back by dropping face ends under back pick (shown by '0') and raising of back end over face pick (shown by '+').

g) Extraction of stitching weave by first method (dropping of face under the back).

h) Extraction of stitching weave by second method (raise of back ends over face pick).

i) Denting plan for the design shown at 'e' (two ends per dent or RTP in a dent).

j) Drafting arrangement with all face ends at front and back ends at back.

k) Peg or lifting plan for the design shown at 'e'.

Example 2: Figures 3.5 l to q

l) face weave 2/1 Z twill on 3 × 3

m) back weave 2/1 Z twill on 3 × 3

n) developed design with 1:1 in face and 1:1 in back weaves, lifts for all face ends on back picks, stitching by both methods

o) denting plan for design 'n' equal to RTP per dent

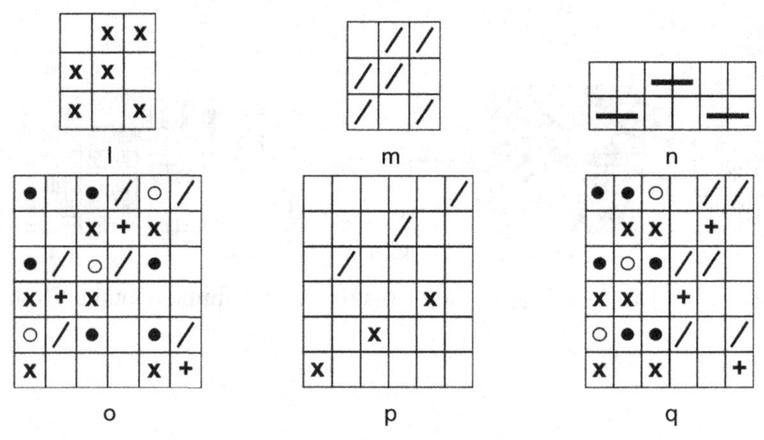

Figure 3.5 (l–q) Self Stitched Double cloths

p) drafting for face and back ends

q) peg plan for the design 'n'

3.4 Self-stitched double cloth with different combinations

In this class of double clothes, we have selected different face and back weave, modification of basic back weave for the purpose of locating binding points, changing the RTP, and so on. The reader is instructed to go through all the examples and shall attend the incomplete work in all these designs.

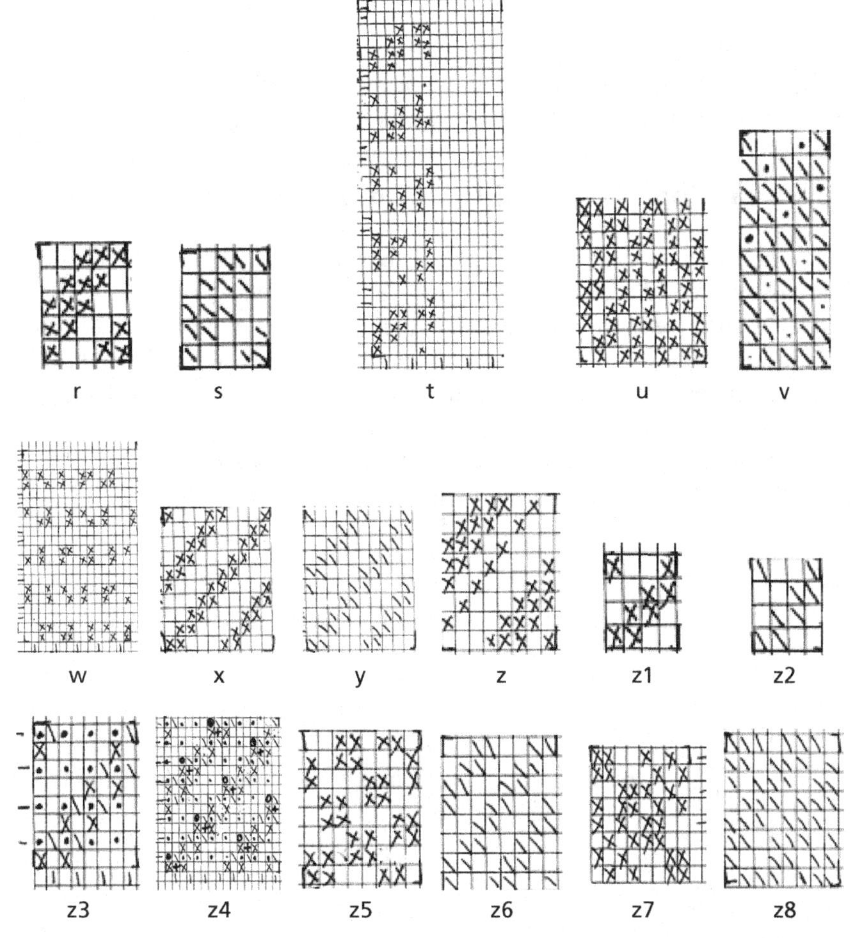

Figure 3.5 (r–z8) Self Stitched Double cloths with different combinations

Figures 3.5r to z8 show various combinations of face and back weave to develop self-stitched double clothes. The examples in Figures 3.5r and s show the 5-end twill as face and back weave with similar footings and Figure 3.5t is developed on 4:2 RTP in weft and 2:1 in warp. The *reader is directed to complete the design.* Figure 3.5u shows a 5-end twill face shown in two repeats and the back weave is a 5-end satin in two repeats (Figure 3.5v) so that the final design will be with 2:2 in weft and 2:1 in warp. The *reader is directed to complete the design of Figure 3.5w.* Designs as face and back shown in Figures 3.5x and y are those which can be produced with only one beam. *The reader is directed to develop and complete the design.* The example of Figure 3.5z is used as both face and back and *the design is to be completed by the reader in all the respects.* Figures 3.5z1 and z2 show the similar weaves on 4-end twill and Figure 3.5 z3 shows the repeat with 1:1 arrangement. The same weaves are represented on two repeats in Figure 3.5z4 with 1:1 arrangement only to facilitate the sateen binding method. The reader is advised to construct the self-stitched double clothes using the weaves given in Figures 3.5z5 to z8.

3.5 Beaming and drafting of double clothes

Like other advanced fabrics, double clothes are also drafted and beamed. The most preferred method is to arrange all front healds for face and back healds for back weave or vice versa. Here, it is to be noted that if the drafting includes alternate healds of face and back, it may cause some difficulties to locate and mend the respective threads in the event of end breakage (Fig. 3.6).

			B	Arranging the healds in an alternate fashion
		F		
	B			
F				
			B	Arranging the healds with all face healds at front and back healds at back fashion
		B		
	F			
F				
			F	Arranging the healds with all back healds at front and face healds at back fashion
		F		
	B			
B				

Figure 3.6 Methods of drawing of double clothes

3.6 Reversible double clothes

Double clothes in which face warp floats are followed by back warp floats or face and back weaves are reversible in nature are called as *reversible double clothes*. The construction of reversible double cloth is similar to simple double cloth. Figure 3.6a indicates 2/1 Z twill as face weave and back weave is on 3 × 3 with 1/2 Z twill and Figure 3.3c shows the developed design.

Procedure of construction is as follows:

1. Select the weaves, that is, 2/1 Z twill as face and 1/2 Z twill as back.
2. Calculate the RTP in warp and weft and here it is 1:1.
3. Transfer face weave on face threads and back weave on back threads.
4. Indicate the lifts for all face ends on back pick except at stitching points.
5. Select the binding points by both methods.

These are used in the production of worsted overcoats and reversible bed sheets in the furnishing sector.

3.7 Centre-stitched double clothes

Why self-stitched double clothes are not preferred?

In self-stitching, certain threads are lowered and some are raised to achieve binding between two layers. Even though this concept sounds well theoretically, it has some drawbacks. When face and back are dyed yarns and movement of these relatively results in reduced contrast or light texture or subdued effect. Hence, it is necessary to use a separate stitching thread which connected to both face and back and such a method of stitching is known as *centre-stitched double clothes*. Centre stitching may be achieved in warp as well as weft direction. The reader is advised to construct the centre warp stitched and centre weft stitched double clothes using the weaves of Figures 3.5z5 to z8.

3.8 Centre warp stitched double clothes

Figure 3.8d shows face weave 2/2 matt on 4 × 4 and Figure 3.8e shows 2/2 Z twill as back weave. The developed design is in Figure 3.8f where *S* stands for centre thread and is connected once to face by '+' mark and connected to back by '0' mark.

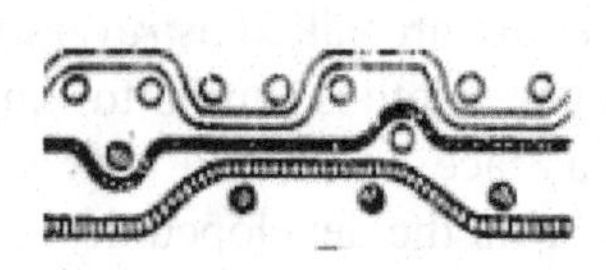
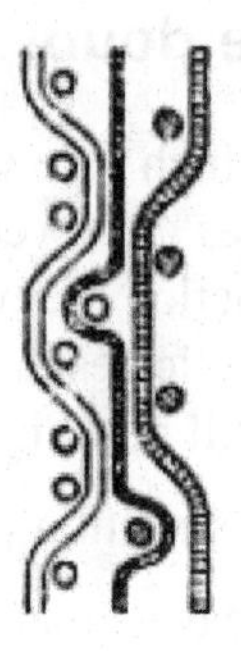

Figure 3.7 Principle of centre-stitched double clothes

Notations used for Figure 3.8f

- X face weave up
- / back weave up
- . lifts for all face ends on back picks and centre ends on back picks (except at tying points for the later)
- O stitching for centre warp by lowering it under the back pick
- + raise the centre end over the face pick

Where should be the point of contact for centre in centre-stitched double cloth?

1. The centre end is lowered on face pick and raised over the back pick when no tying occurs.
2. In tying centre end to face, it is accommodated between two adjacent face warp float and is raised over the face pick and hence these must be away from the face side.
3. In tying centre end to back, it is accommodated between two adjacent back warp float and is lowered under the back pick where it is away from backside.
4. The centre end is raised over the face pick and lowered under the back pick. The centre thread may be disposed anywhere in the repeat. Minimum of one thread and a maximum of three threads are recommended.

Following is the procedure of construction: Select the face and back weaves, calculate the RTP and mark the final repeat size and mark RTP, transfer face weave and back weave on respective threads, lift all the face ends on back

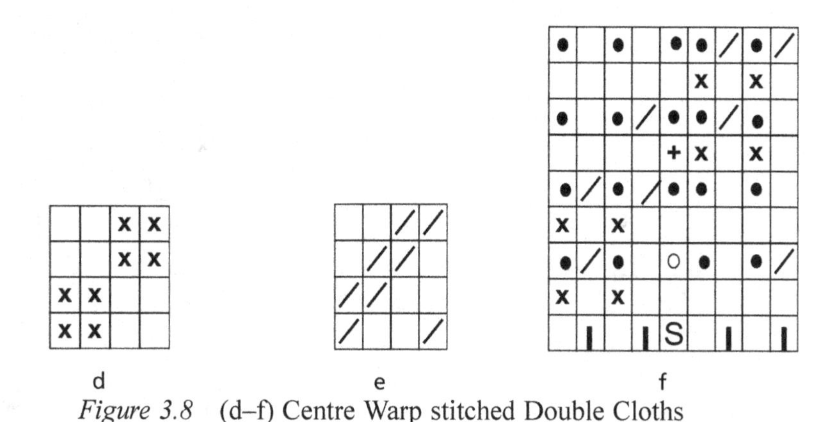

Figure 3.8 (d–f) Centre Warp stitched Double Cloths

picks for the formation of top layer of shed, connect the centre thread once with face pick and one with back pick. The weaving arrangement consists of drawing stitching end in a separate heald shaft. The lifting plan indicate the lifts for centre threads on all back picks except at the tying points along with lifts for face ends on back picks. A small double-flanged bobbin of 12 inches diameter or any other suitable small beam is arranged at the back side of loom. The count of centre thread may be compatible with others otherwise the structure may look like use of wadding thread.

Tartan-line overcasting

It is produced from finer yarns and is woven face down in weaving. Details are as follows:

Face weave – 2/2 Z twill on 4 × 4, back weave – 2/2 Z twill on 4 × 4, centre threads – 4.

Arrangement FBFSB four times in warp and 1: 1 in weft.

3.9 Centre weft stitched double clothes

The construction of centre weft-stitched double cloth is similar to centre warp-stitched except the way of connecting centre weft to face and back layers. In centre weft construction, once face end is dropped under the centre pick and once back end is raised over centre pick. Generally, during centre weft pick insertion, all face ends are lifted and back ends are lowered except at the connecting points of centre with face and back. Figure 3.9g shows face weave, Figure 3.9h give back weave, and Figure 3.9i is developed design with

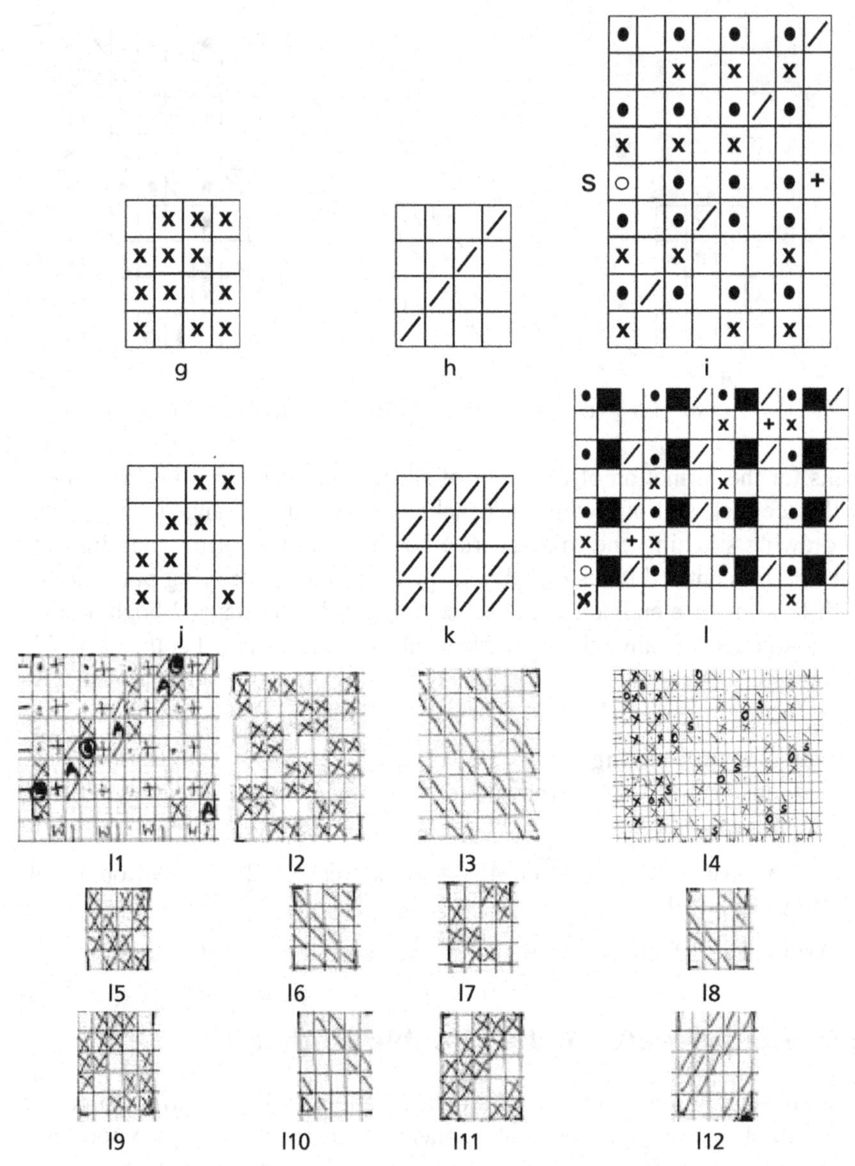

Figure 3.9 (g–l12) Centre Weft Stitched Double Cloth

all necessary details. In the design, 'O' indicates the lowering of face end under the centre pick and '+' indicates the lift of back end over centre pick.

The reader should note that a group of weaves are given in Figures 3.9l5 to l12 and double clothes should be constructed using these weaves.

Notations used for Figure 3.9i

X face weave up

/ back weave up

. lifts for all face ends on back picks and centre picks (except at tying points for the later)

O stitching for centre weft by lowering the face end under the centre pick

+ raise the back end over the centre pick

Further, in Figures 3.9l1 and 14, letters 'A' and 'S' stand for stitching by the second method. *The reader must note this in notation.* The reader should also note that the wadding end is lifted on all back picks by indicating '+' or 'X' in these designs.

Why centre weft stitching is not preferred over centre warp stitching?

Centre weft stitching reduces the rate of production due to the fact that when centre weft picks are introduced, the take-up motion should be made inoperative and thus the process does not contribute to the length of fabric produced.

Following is the procedure of construction selecting the stitching point

Regular construction

1. Select the face and back weaves.
2. Select the number of centre weft usage.
3. Mark the final repeat including centre thread.
4. Transfer the face and back weaves respectively on the respective threads.
5. Indicate the lifts for all face ends on all back picks and similarly face ends are raised on centre weft picks.

Stitching point selection

- Drop the face end under the centre pick at a point where it is absent from the surface and covered by two adjacent floats of face weft.
- Raise the back end over the centre pick at a point where it is absent from the back side and covered by two adjacent float of back weft.

Weaving arrangements

Use separate shuttles for face, back and centre picks. Give the instructions for peg plan as to peg or cut all the marks except blanks and 'O'. Use separate healds for face and back ends.

When do you prefer centre weft stitching?

1. When all the jacks or harness are entirely used for face and back and none are available for centre warp threads.
2. When conditions do not encourage the use of a separate beam carrying centre warp threads.
3. When control or accessibility of centre warp threads is difficult

3.10 Wadded double clothes

The role of wadding thread is already discussed in previous chapters and more or less remains the same even in double clothes (Fig. 3.10). Wadding is affected either in warp or weft direction by using a separate heald shaft or separate shuttle. Wadding yarn will be inferior to main yarns so that the cloth is produced economically. Wadding not only increases the weight but also accounts for the appearance of the fabric. When wadding threads are used, the strength of the fabric increases in the direction of usage. The use of wadding threads does not affect the cloth appearance and lie at the centre without being seen either at the face or back. However, in some cases, when conditions warrant, wadding threads are connected to the main structure by binding suitably. The number of wadding ends normally depends on the number of respective threads in the principal direction. For example, face weave 2/2 Z twill and back weave 4-end sateen, that is, if wadding is in warp direction, then four wadding ends are used in warp direction controlled by a separate heald, wadding heald. If wadding is in weft direction, four wadding picks are used and will be inserted by separate wadding shuttle. If the wadding thread

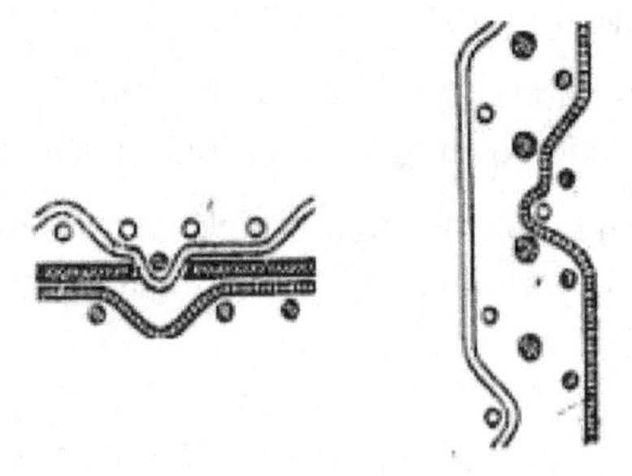

Figure 3.10 Principle of wadded double clothes

used is too thick, then the RTP will be 2 face and 2 back or 2-face–1-back. It should also be noted that the construction of wadded double cloth is not different from regular self-stitched double cloth as wadding is merely made to lie at centre. But necessary instructions are to be given while preparing lifting plan for pegging or card cutting.

3.10.1 Warp wadded double clothes

Wadding thread is in warp direction and it is found that wadding thread can be used more conveniently and economically in warp direction than in weft direction. But sometimes, greater strain is found on the wadding end. Wadding ends will lie at the centre and is obtained by lifting all the wadding ends on back picks and lowering the wadding ends on all face picks. Figure 3.9j shows the face weave on 4 ends and picks of 2/2 Z twill, Figure 3.9k shows the 3/1 Z twill on 4 ends and picks, and 'Figure 3.9l' shows the development of the design with 4 wadding ends and 1:1 in warp and weft.

Notations used for Figure 3.9l

X face weave up

/ back weave up

. lifts for all face ends on back picks

O stitching by first method (dropping of face end under the back pick)

+ stitching by second method (raising of back end over face pick)

Solid mark – Lift all the wadding ends on all back pick

Weaving arrangement

Use of box looms is recommended as separate shuttle carrying face and back pick is needed. Care is to be taken in preparing the box motion chain to avoid the three box motion. Other conditions remain as usual like use of separate healds for face, wadding, and back and drawing the RTP in warp in a dent and arranging the face, back, and wadding healds.

3.10.2 Weft wadded double clothes

Wadding picks selected do depend on the number of back pick and nature of the back pick in relation to wadding pick. An example of weft wadded is shown in 3.11d1 with face and back weave as 3/2 matt with 5 wadding picks. Figure 3.11m shows 2/2 Z twill as face and back weave, Figure 3.11o shows mounting of threads with wadding and show the complete design. This design shows four wadding picks as the face picks are four. The reader is instructed to complete the weavers' plans along with loom equipment and card cutting

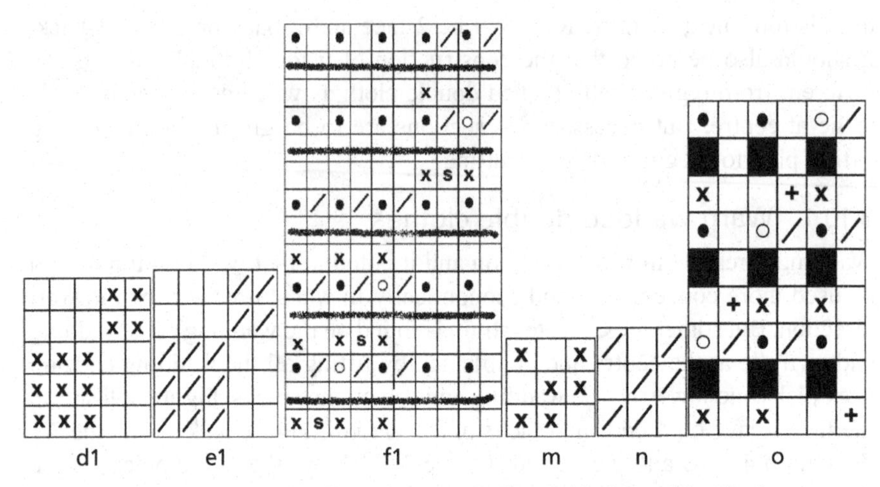

Figure 3.11 (d1–f1 and m–o) Weft Wadded Double Cloth

and casting out instructions if ends per inch are 50 and picks per inch is 60 for a design repeating on 7 inches width and 7 inches in length.

Notations used for Figure 3.11o

X face weave up

/ back weave up

. lifts for all face ends on back picks

O stitching by first method (dropping of face end under the back pick)

\+ stitching by second method (raising of back end over face pick)

Solid mark – lift all the face ends on all wadding pick

Notations used for Fig. 3.11f1

X face weave up

/ back weave up

. lifts for all face ends on back picks

O stitching by first method (dropping of face end under the back pick)

S stitching by second method (raising of back end over face pick)

\+ lift all the face ends on all wadding pick

Solid horizontal line – lift all the face ends on wadding picks

Following is the constructional steps in general

1. Select the face and back weave.
2. Select the wadding picks based on the back picks number.
3. Find the repeat size and mark the RTP in warp and weft, wherein weft side includes the wadding picks.
4. Transfer the face and back weaves.
5. Indicate the lifts for all face ends on back picks except at the stitching points (first method).
6. Lift all the face ends on all wadding picks and drop all the back ends on wadding picks so as to make it lie at centre.
7. Locate suitable positions for stitching by first and second method to complete the design.

Weaving arrangement
Use of box looms is recommended as separate shuttle carrying wadding pick is needed. Care is to be taken in preparing the box motion chain to avoid the three-box motion. Other conditions remain as usual like use of separate healds for face and back and drawing the RTP in warp in a dent and arranging the face and back healds in any one of the fashions as mentioned with respect to beaming and drafting of double clothes.

3.11 Interchangeable (IC) double clothes

It has been mentioned at the beginning of the chapter that double cloth can be classed as *interchangeable double cloth* when the interchanging of the layers occurs, that is, face layer will work as back and vice versa at some point which is known as binding or concealing point. These are very popular in decentralized sector as colour contrast is brought significantly as compared to self-stitched double cloth. The main advantage is due to the fact that the layers interchange rather than one or two threads interchange. Another reason for popularity is that they are produced using an inverted hook jacquard wherein the movement of the card is used in both the directions. Due to these reasons, interchanging double clothes are preferred than simple double clothes.

3.11.1 Types of threads and RTP in IC double clothes

Similar to the regular double clothes, IC clothes have two series of threads in warp and weft, namely dark and light, which may be disposed in different RTP. Generally, it is 1:1 in warp and weft directions.

Procedure for construction of interchangeable double clothes:

1. Consider the pattern which contains dark and light portions and this may be any geometric portion or a small motif or spot effect.
2. Based on the size of the above with 1:1 in warp and weft for dark: light, work out the number of final repeat and mark the RTP in warp and weft.
3. First mark one down and one up plain on dark ends with 'x' and work the next dark end with one up and one down. Similarly, work out plain for light end with '/' mark.
4. Refer the basic pattern for dark or light portions.
5. For any dark portions, it is necessary to lift all dark ends on light picks and is to be indicated by dots.
6. Similarly, for any light portion, indicate the lifts for all light ends on all dark picks which completes the design.

3.11.2 Systematic construction of IC double clothes

Figure 3.12a shows a portion of design to be developed for IC double clothes. Here, the portion is a geometric figure and has two picks and four ends, that is, totally it has four picks and eight ends (dark + light in 1:1). Mark the repeat size for four picks and eight ends. Mark RTP as one dark and one light. All light threads are indicated by 'dash' 'l'. Mark the plain weave for all dark and light ends as explained in the general procedure. Count the number of picks and ends of dark portion and here the first dark portion has two picks and

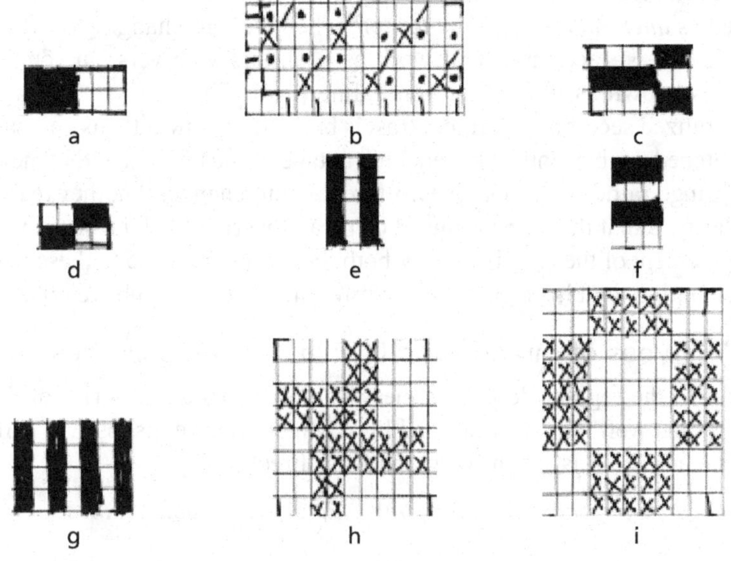

Figure 3.12 (a–i) Systematic construction of IC double clothes

two ends. Therefore, indicate lift for first two dark ends on two light picks by 'dots' and now refer the light portion. There are two picks and two ends for light. Indicate the lifts for third and fourth light ends (as first and second light ends are on the bottom side) on dark picks (i.e. on first and second dark picks). Figure 3.12b shows the developed design. Different patterns are shown in Figures 3.12c to g and the reader is directed to develop with 1:1 RTP.

Drawing and denting arrangements

All dark ends are drawn through dark healds placed either at the front or back. All light ends are drawn through light healds placed either at the front or back. RTP should be drawn per dent.

Card cutting instructions

Cut all the marks except blanks.

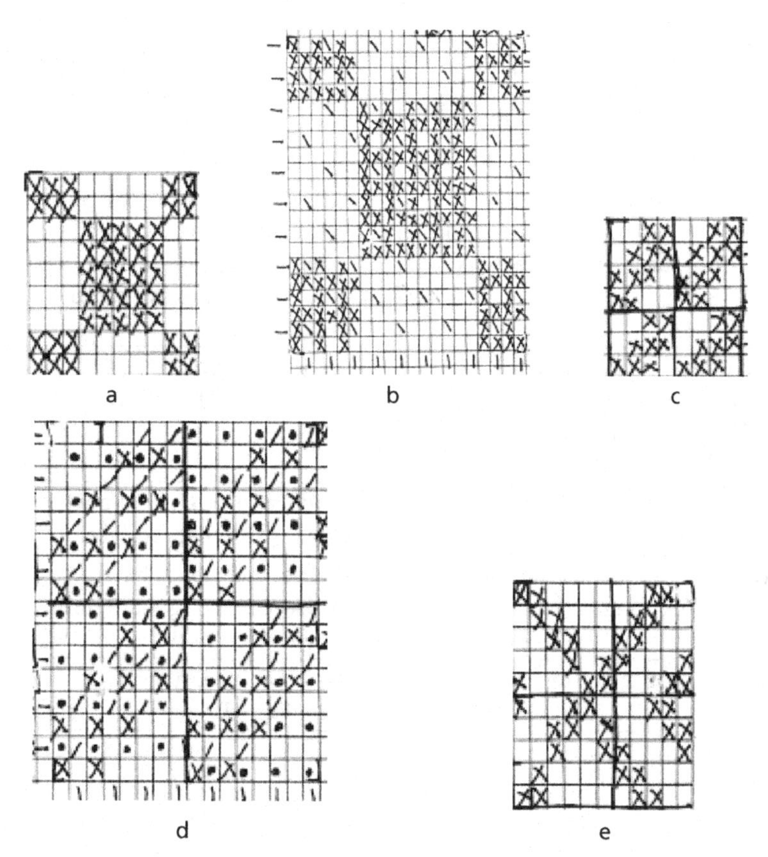

Figure 3.13 (a–e) Cut double clothes

3.12 Cut double clothes

These are the double clothes known as cut double clothes as thread (part) will interchange rather than the entire layer as observed in IC. Figure 3.13a shows the motif and is developed in Figure 3.13b. A similar example is shown in Figure 3.13c and development in Figure 3.13d. The design at Figure 3.13e is for the reader to complete.

3.13 Differences between backed clothes and double clothes

Backed fabrics	Double clothes
1. It is a semi-compound fabric.	1. It is a compound fabric.
2. It consists of two series warp and one series weft or two series weft and one series warp.	2. It consists of two series of warp and two series of weft.
3. Warp-backed fabrics require two beams and extra harness, while weft-backed fabrics require drop box looms	3. They require, two beams, extra harness, and drop box motions
4. Stitching is done either by dropping face end on back pick or lifting back end on face pick.	4. Stitching may be accomplished by three methods: (1) self-stitching, (2) centre stitching, and (3) stitching by thread interchange.
5. Wadding thread may be introduced in only one direction, either warp (for weft backed) or weft (for warp backed).	5. Wadding threads may be introduced either in warp or weft.
6. It is cheaper to produce than double clothes.	6. Production cost is more when compared to backed fabrics.
7. Weave gives scope only for inferior figuring effects.	7. Weave lends itself for the development of good figures on the face and back.
8. Designing the cloth is simple.	8. Designing the fabric is complicated.

CHAPTER 4

Treble clothes

4.1 Introduction

Treble clothes form three series of warp and weft threads which form three distinct fabrics, one above the other. Except for the ties, when a face pick is inserted, all the centre and back ends are left down; when a centre pick is inserted all the face ends are raised, and all the back ends are left down; while when a back pick is inserted, all the face and centre ends are raised. The face ends and face picks interweave to form the face fabric, the centre ends and centre picks weave to form the centre fabric, and the back ends and picks constitute the back fabric. The three fabrics are joined together by interweaving the centre ends and picks with the face and back ends and picks, and the resulting cloth has firmness and weight equalling that of the three single fabrics. Thus, greater weight than the double system is possible, and at the same time, cloth of same fineness of appearance can be produced. Sometimes, the weight of double woollen structures is increased by shrinking the cloth in the milling process, but this affects the elasticity, air permeability, and clarity of effect. But all this can be avoided and at the same time increased weight can be obtained if the treble principles are employed.

4.2 Methods of stitching

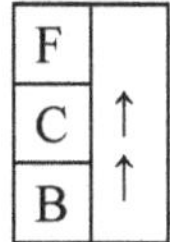

1. Centre ends are lifted on the face picks and the back ends on the centre picks. This is the most common method employed.

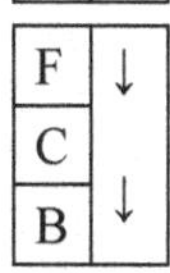

2. Stitching can be affected by dropping the face ends under the centre picks, and the centre ends under the back picks. In each instance, therefore, the typing consists of the cancellation of certain selected separating lifts.

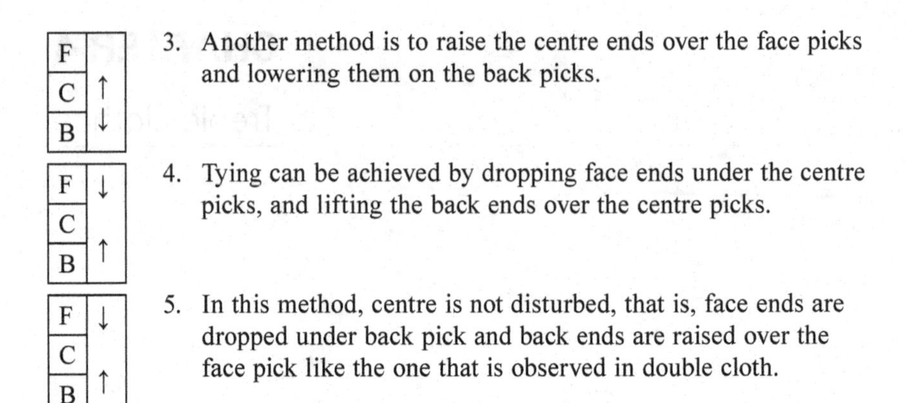

3. Another method is to raise the centre ends over the face picks and lowering them on the back picks.
4. Tying can be achieved by dropping face ends under the centre picks, and lifting the back ends over the centre picks.
5. In this method, centre is not disturbed, that is, face ends are dropped under back pick and back ends are raised over the face pick like the one that is observed in double cloth.

Although the first method is the most popular (Fig. 4.1), the other three methods may be preferentially adopted when the positions of convenient binding points in the face and back weaves are unsuitably placed if the first

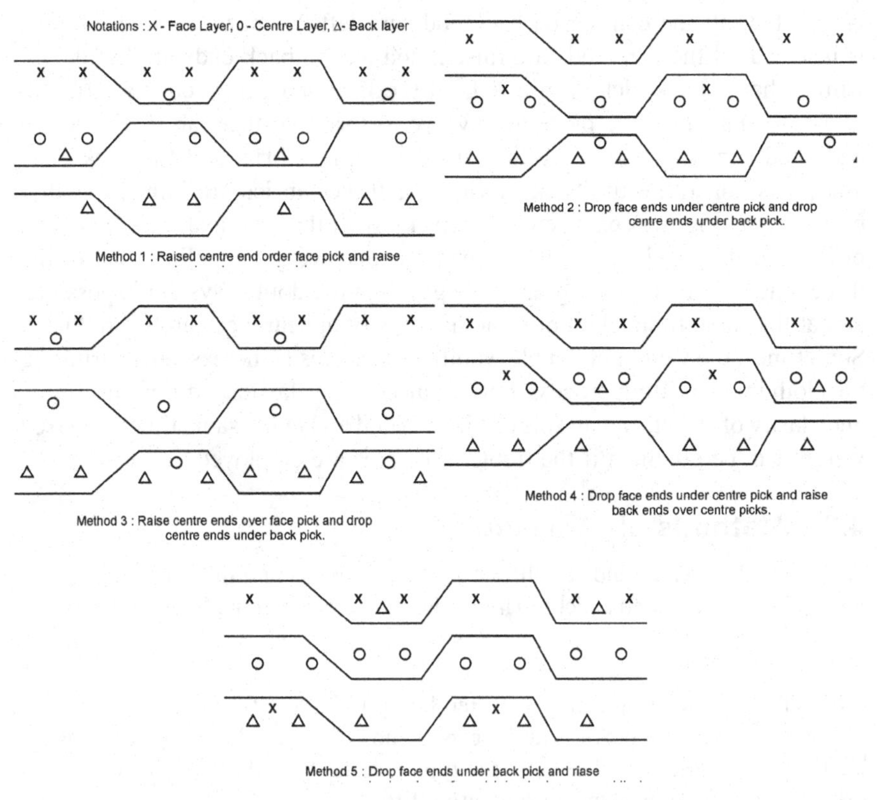

Figure 4.1 (a–e) Methods of stitching in treble clothes

method is employed; or, if relative thickness or quality of the yarns is such that the other methods result in a lesser degree of disturbance to the visible surface of the cloth. As the centre is never visible, the position of the tie in that layer is of no particular importance in itself; however, efforts should be made to adhere, if possible, to the normal rules of stitching even in respect of the centre layer because any undue disturbance of its regularity is liable to result in a degree of distortion in the face or the back layer. Apart from the above-mentioned four methods of stitching, it is also possible to tie treble cloth together by either dropping face ends on back picks, or, lifting the back ends on face picks, as in an ordinary double cloth. In this method of stitching, the centre cloth acts merely as a welding layer.

4.3 Construction of treble clothes

When identical weaves are required in all layers, the weave in each of the three fabrics – face, centre, back – is the same and they are all started on the same footing. This also creates the most favourable condition for tying. Once the weave of each fabric layer is determined, the construction can be commenced in the following order:

1. The order of arrangement of the three series of threads is determined and indicated around the margin of the design repeat.
2. Insert face weave on face ends and picks, the centre weave on centre ends and picks and the back weave on back ends and picks. Different marks are used to denote the three different pick weaves to avoid confusion.
3. Insert the separating lifts. This is achieved by lifting all the face ends on all the centre and back picks, and by lifting all the centre ends on all the back picks. It follows that on the face picks, all the centre and back ends must be down, and on the centre picks, all the back ends must be down.
4. Introduce stitching marks, the treble clothes are usually self-stitched and in the same way as the double clothes are stitched. It is important to remember the following points when stitching:
    - The stitching should be as regular as possible to prevent uneven tension.
    - The ties should be adequately concealed on the face and on the back by corresponding face and back floats.
    - The face and the back ends can only be used for stitching purposes when absent from the visible surfaces of their respective clothes.

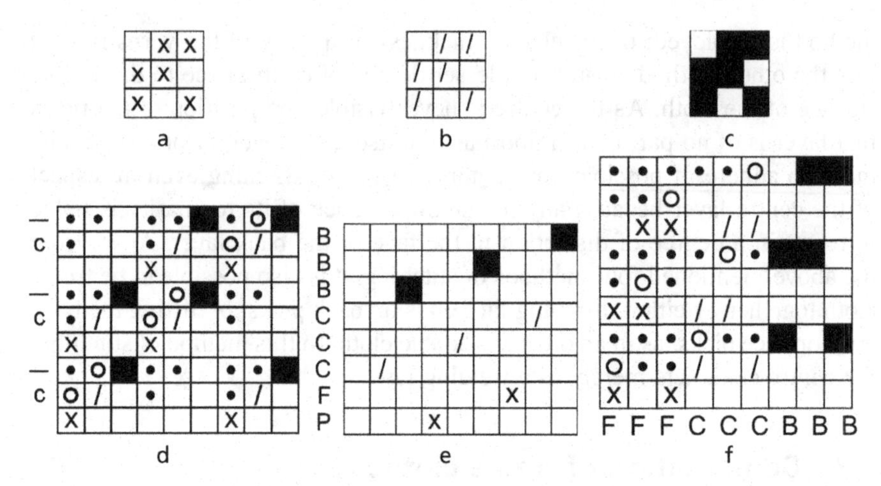

Figure 4.2 (a–f) Simple treble cloths

Figures 4.2a, b, and c show 3-end twill on 2/1 as face, centre, and back weaves with 9 × 9 final size as shown in Figure 4.2d. Drafting arrangement is shown in Figure 4.2e and Figure 4.2f gives the lifting plan.

Notations for Figure 4.2d

X face weave up,

/ centre weave up

Solid mark – back warp up

. lifts for all face ends on back and centre picks and centre ends on back picks (except at tying points)

O stitching – dropping face end under centre pick

D stitching – dropping centre under back pick

4.4 Beaming of treble clothes

It is always advantageous from the point of view of the preparation of warp and access to the weaving machine to have a single beam. But this is possible only when the material and the weaves of the three layers are similar. When two layers are similarly constructed, two beams will have to be used, but when all the three layers are dissimilar or when the take-up in each layer varies it is necessary to use three beams.

4.5 Drafting of treble clothes

Different ways are available for drafting treble clothes. Various arrangements are possible and few are shown below and further ways are also possible. But from the working point of view, it is better to have either all face at front, centre at middle, and all back at back. The method of drafting to be employed is dependent on its regularity and the ease with which the ends are drawn-in both during the entering and also following the breaks in weaving. While straight draft can be employed, it is usual to separate the healds into sets to correspond with each fabric layer. This method helps in easy checking of designs and pattern chains. But the selection of draft, whether straight or otherwise should fulfil the primary requirement of regular end distribution. Whether the face, centre, back clothes sets are in the front, middle, or back of the weaving machine is a matter of choice. However, if the geometry of the weaving shed formation in a weaving machine is such that the front healds cause the least strain in the yarn then, undoubtedly, the weakest yarns, or, the most crowded and most frequently interlacing sets of healds should be placed in front.

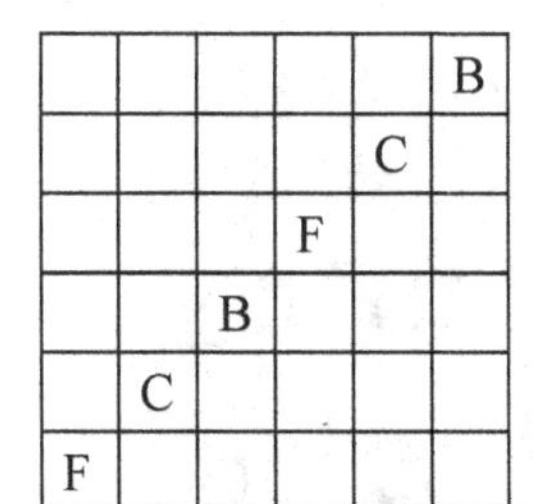

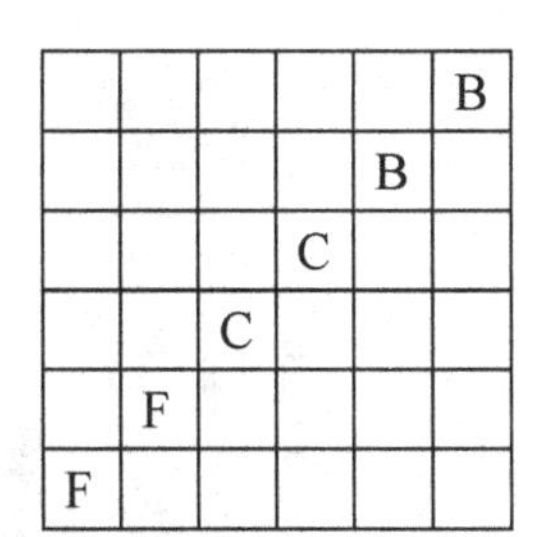

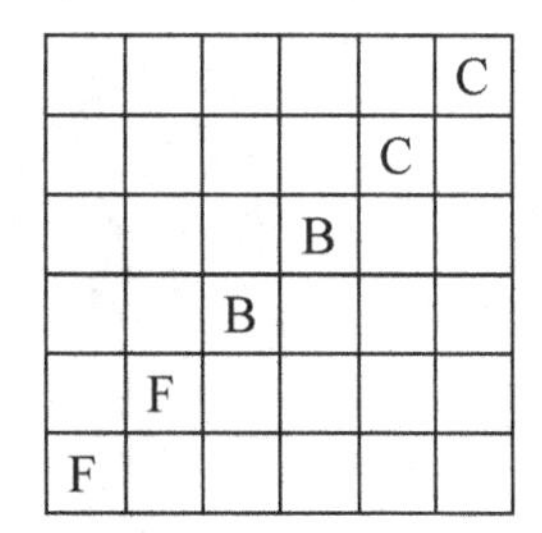

4.6 Treble clothes with dissimilar weaves in the different fabric layers (Fig. 4.3 to Fig. 4.6)

While explaining the method of construction of treble clothes earlier, it was assumed that all layers have the same weave, the threads are arranged in equal proportions, and all the weaves are commenced from the same relative position so that tying can be easily be done. When different weaves are employed in the different layers, it is not possible to get tying positions as easily as before, and hence it may be necessary to experiment a bit before the best relative starting positions for each weave is determined. The construction of fully worked-out design should not be commenced; therefore, each of the constituent weaves is arranged separately in such a relationship with one another that the desirable coincidence of the warp and weft floats is obtained. The weaves are usually marked on transparent design paper and superimposed in such a way that the

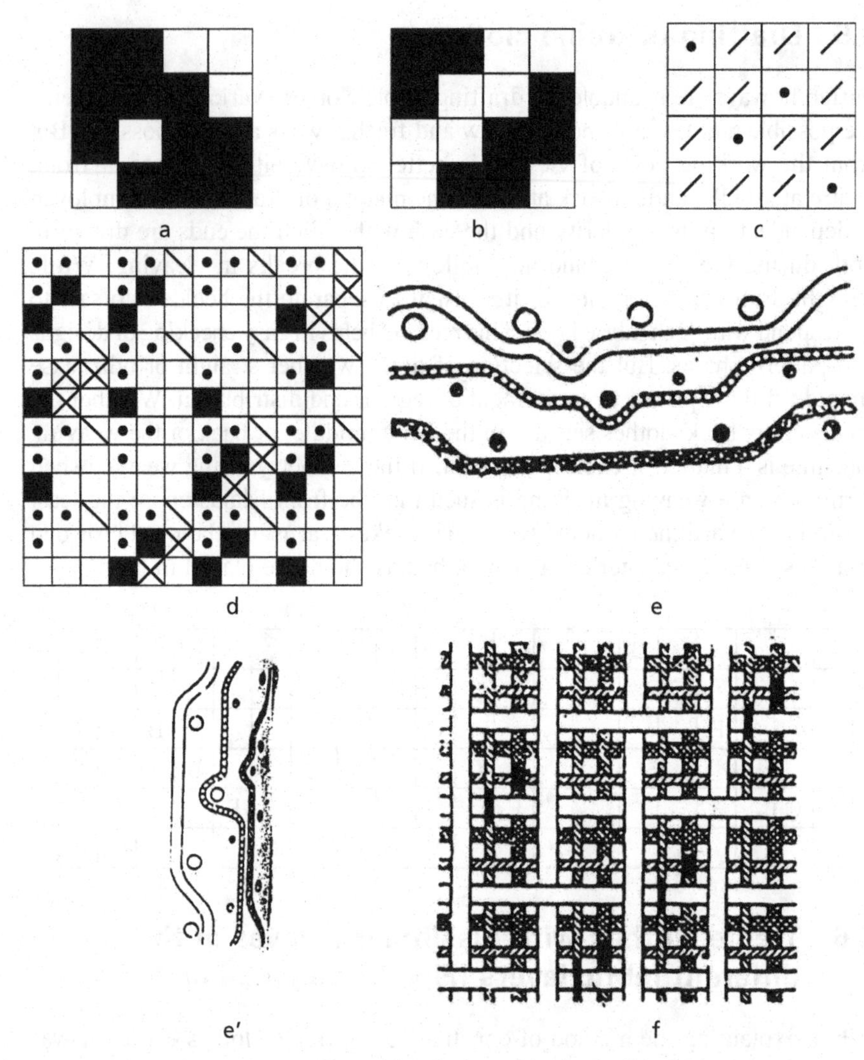

Figure 4.3 (a–f) Treble cloths with dissimilar weaves

greatest possible degree of coincidence between the warp-on-warp, and weft-on-weft float is achieved. This, in fact, is most easily done in pairs because the two relevant relationships from the point of view of stitching are given below.

a) The position of the face weave in respect to the centre weave and

b) The position of the centre weave in respect to the back weave.

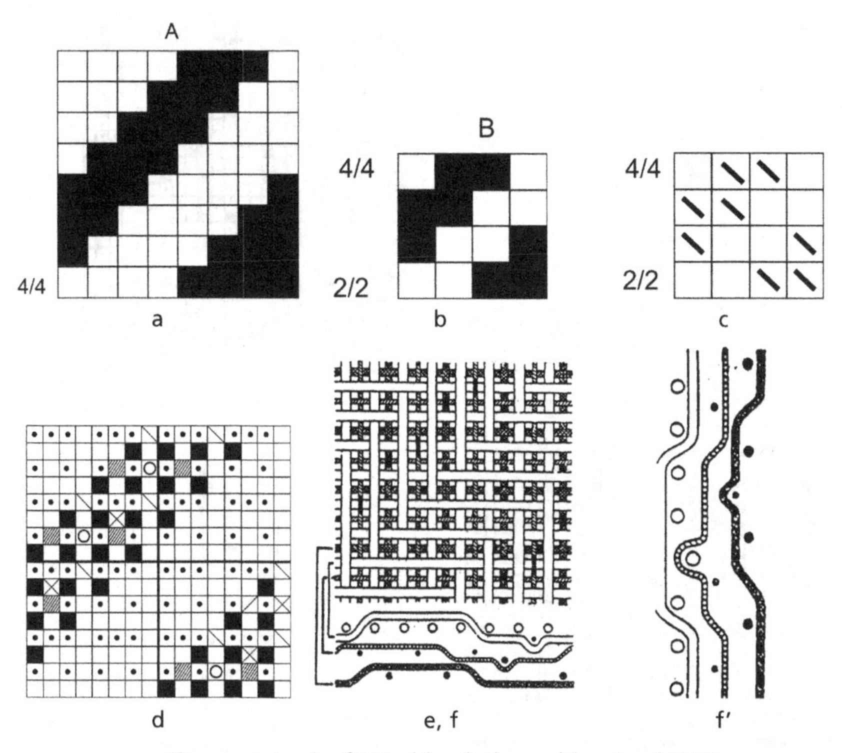

Figure 4.4 (a–f′) Treble clothes with mixed RTP

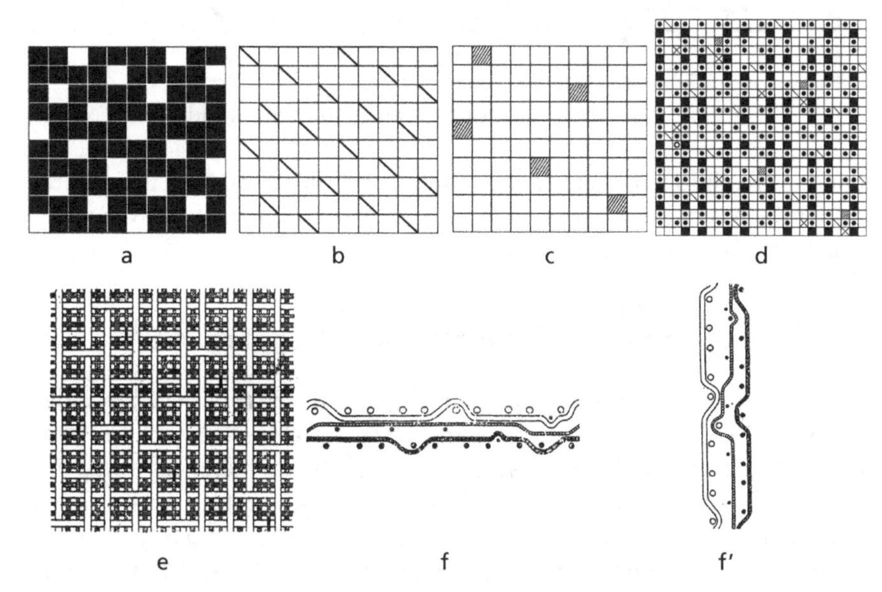

Figure 4.5 (a–f′) Reversible Treble Cloths

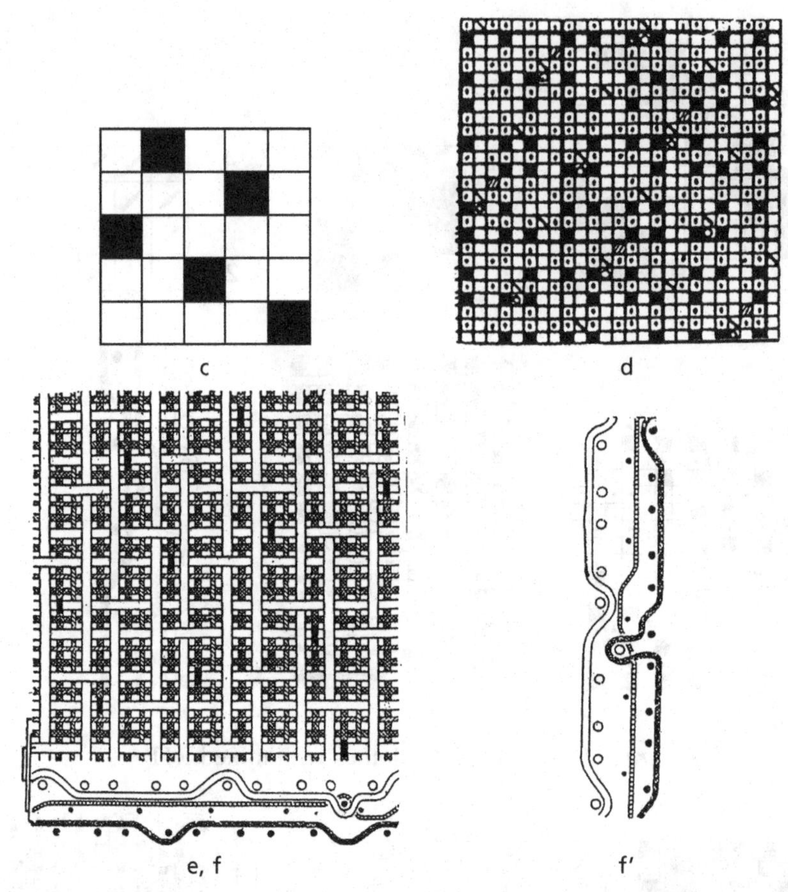

Figure 4.6 (c–f') Use of the centre layer as wadding

It must be realized that should it be entirely impossible to produce the required relationships between the pairs of weaves the weave of the centre layer, which is never visible could be modified.

4.7 Use of the centre layer as wadding (Fig. 4.6)

It has earlier been mentioned how it is possible to tie the treble cloth by either dropping face end on back side, or lifting the back ends on face picks, as in an ordinary double cloth. In this case, the centre cloth acts merely as a wadding layer. This system is advantageous in the sense that centre yarns of lower quality and greater thickness can be used. In arranging the positions of the weaves and ties, it is only necessary to consider the face and back fabrics, as in the case of the self-stitched double clothes.

CHAPTER 5

Pile fabrics (terry pile)

5.1 Introduction

Terry or Turkish towels were originated in Constantinople, Turkey, wherein these fabrics were woven in handlooms. In the middle of the 19th century, this technique of weaving towels was further refined in the European countries and took a shape of power-driven looms (Hobson 1990). Terry towel developed in Turkey were later groomed as towel manufacturing units spreading across the world. Terry fabrics basically classed as pile fabrics, wherein an additional yarn used to form a loop or pile giving a distinct appearance. Pile means 'loop', fabrics with projecting loops fabrics are called as pile fabrics. The loops may be cut or uncut. Pile weaving is another method of fabric structure in which a body cloth is woven and extra threads of weft or warp, employed to give either more thickness or more ornamentation. The characteristic feature of these weaves is the formation of a series of loops (∩) projecting from the main body of the fabric. These loops are formed by (Kienbaum 1978, Ramaswamy 1992) the extra series of threads (warp and weft) and may be distributed uniformly either on the face side only or on both the face and back of the fabric to form a perfectly even surface, or the loops of pile may be developed in such a manner as to create a figured design on a plain or simple ground.

5.2 Classification of pile fabrics

The pile fabrics may be classified in various ways as follows:

1. Warp pile and weft pile fabrics: If the pile or loop is formed by the warp yarns, then the fabric is termed warp pile fabrics if the loops are formed by the weft, they are termed weft pile fabrics.
2. Cut and uncut pile: If the loops formed due to the weave are cut in a subsequent process, such that short tufts of fibres are made to stand

erect, then such fabrics are referred to as cut pile fabrics. The way the cutting is done is also one of the criteria for classification if the cutting is done uniformly over the entire face of the cloth then such fabrics are termed as velvets or velveteens. But if the cutting is done in such a manner that the tufts of fibre formed to give a ribbed appearance, then such fabrics are called corduroys. If on the other hand, the pile is left as it is, then such fabrics are called uncut piled fabrics.

5.3 Terry pile fabrics

The name 'terry' comes from the French word 'tirer', which means to pull out, referring to the pile loops which are pulled out by hand to make absorbent. Actually, a research was conducted on terry weaving by Manchester Textile Institute in which the terry weaving was concluded as defective weaving. The original research occurred in Turkey in Bursa City and later terry weaving was considered as a development of woven fabrics.

What are terry pile fabrics?

The terry pile is a class of warp pile structure in which certain warp ends are projected from the ground surface in the form of loops (Figs. 5.1 and 5.2). The fabrics are produced basically in 100% cotton as they are used for moisture absorption. However, other than moisture absorption fibres such as viscose, acrylic are considered. The fabrics are produced on looms provided with what is known as 'terry motion.' The width of the fabric produced depends on the type of application. Fabrics may be produced only in plane form (only piles without any figures), or stripes or checks or figured or all over patterns or

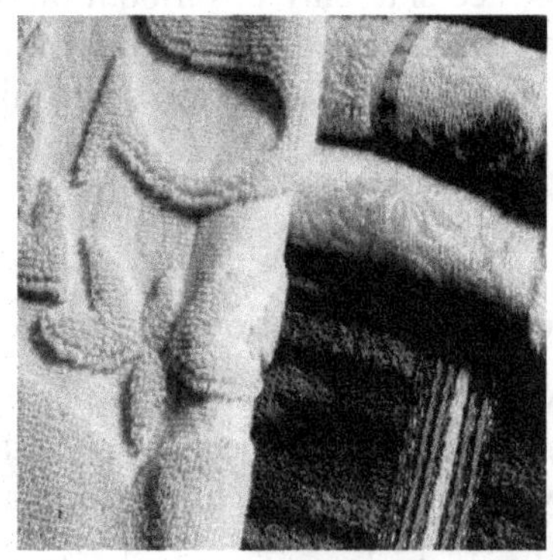

Figure 5.1 Jacquard design terry towel

single large figure like tapestry type. No doubt ornamentation is fully brought using multicoloured threads.

What is RTP of terry pile?

Terry belongs to complex or advanced fabric structure in which weight of the fabric is increased by 1.5 times as extra threads are used only in the warp direction. Only one series of weft threads are used but the warp consists of two series namely ground and pile. The former produces with the weft the ground cloth from which the loops formed by the pile ends project. The loops may be formed on one side only or on both sides of the cloth; thus, producing single-sided and double-sided structures, respectively. Accordingly, for single-sided structures, we have ground warp and pile warp in 2:1 ratio and for double-sided designs ground warp, face pile, back pile in 2:1:1 ratio.

Figure 5.2 Intermittent type terry

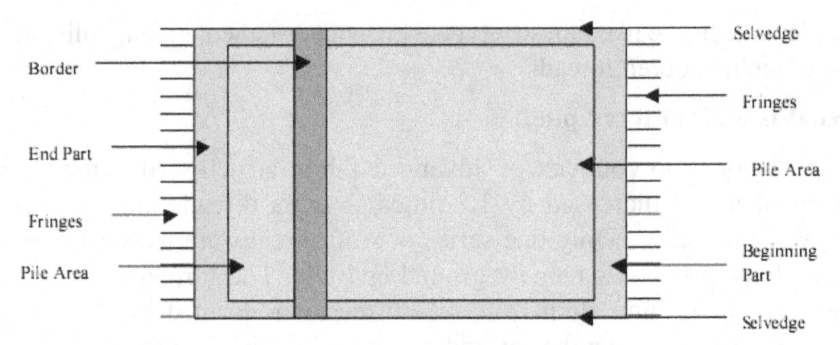

Figure 5.3 Various parts of a typical terry

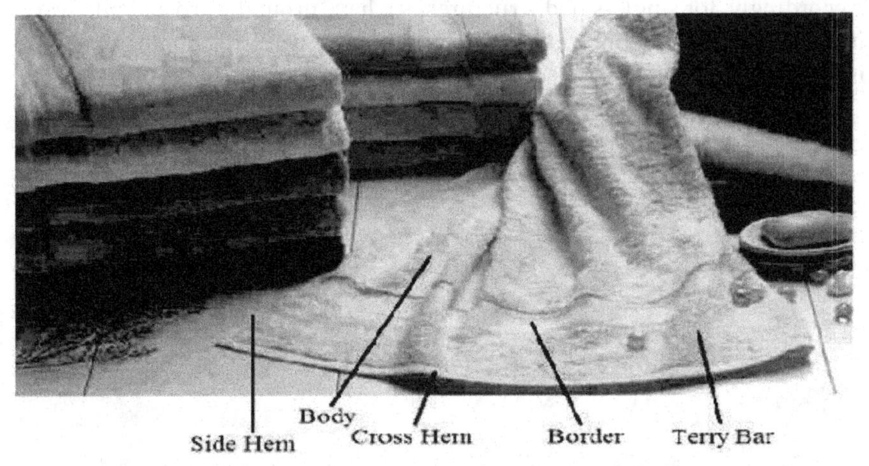

Figure 5.4 Various parts of a commercial terry

5.4 History, research, and development of terry fabric production

5.4.1 Introduction

Kienbaum (1978) described various production techniques, construction, and patterning range of terry towels. He observed that a variety of methods can be employed to manufacture terry towels (but among these weaving, knitting, sew knitting, and needle pile methods are important). Each of these methods contributes a different range of functional properties and esthetic appearance. Victor Hobson (1990), in his work, described various mechanisms of terry loom being developed at the initial stage. He observes that initially handlooms were used to manufacture terry towels, with the aid of long rods in weft direction so that the height or size of loops are varied based on rods dimension. Later, terries were produced on power looms equipped with mechanical means.

At the end of 19th century, the researcher notes that all the accessories of present-day terry were developed.

5.4.2 Research studies on terry towels

Following paragraphs give the bird eye view about research work reported globally on terry towels.

Swani et al. (1984)

The study included the production of terry towels by using the ring and the OE (Open End) yarn in pile warp and studied the functional properties. It was observed that among various parameters of yarn, linear density, twist per inch, and packing density play a significant role in achieving the desired quality of terry towels. The higher packing density of yarn resists absorbing more quantity of water due to the availability of lesser space to retain water. Further, he observed that there was no difference in water absorption rate in towels made out of OE yarn and ring yarn but the maximum absorption of water for OE towels was better than that of ring towels. They found that there was no significant difference in the water absorption rate of ring and OE fabrics although yarn wicking rate was greater in fabrics of OE yarn. Absorption rate increased with increase in pile density but was not affected by pile ratio. Further, he added that maximum absorption for OE fabrics was better than for ring fabrics at lower fabric density and for comparable fabric weight. Ring fabrics showed better dry and wet abrasion resistance than OE fabrics. The abrasion resistance of the fabrics was maximum for moderate pile density and high pile ratio. The wet abrasion resistance was significantly lower than dry abrasion resistance of the fabrics.

Ishtiaque (1986a, 1986b)

In his study, he observed that the packing density of fibres in ring yarn increases with the increase of twist. Various researchers observed that the towels manufactured with the low-twist yarn in pile warp have better functional properties in terms of softness and water absorbency.

Ramaswamy (1992)

He compared the three-pick terry and four-pick terry and observed that four-pick terry fabrics are heavy in structure and provide better quality in comparison to three-pick fabrics.

Mukhopadhyay et al. (1996, 1998)

The study included the preparation of a bi-component yarn by blending water-soluble PVA (Polyvinyle Alcohol) staple fibre with cotton fibre and used this yarn in pile warp for producing terry towels. After completion of mechanical processing of weaving, towels being prepared by using bi-component yarn were treated in boiling water to dissolve the water-soluble PVA component of yarn. As a result of the dissolving of PVA fibres, the remnant yarn was more bulky, which lead to the lower packing density of fibre in yarn. They observed that the towels made out of bi-component yarn were better in terms of various functional properties as compared to the towels of ordinary ring spun yarn. The towel also exhibited improved water absorbency, higher abrasion resistance, softer and lighter weight than towels woven using conventional yarns. The process had no adverse effect on light fastness or wash fastness.

Mansour et al. (1997)

Mansour et al. worked to develop an expert system for terry weave. Most important weaving parameters being identified by them were cut and uncut loops, pile density, and pile height. Manipulation of these weaving parameters affected various physical and mechanical properties such as water absorbency, tensile strength, abrasion resistance, fabric weight, thickness, handle and so forth.

Srivastava et al. (1997, 1998)

They worked on optimization of process parameters to improve functional properties of bi-component terry towel. The process parameters being considered as most influencing were twist in yarn, the proportion of water-soluble fibre (WSF) component, pile ratio, and picks per inch. These parameters decide the terry towels' functional properties such as moisture absorbency, water retention, drying ability, resistance to abrasion, softness, and feel. Results indicated that optimal bi-component towels in respect of above-mentioned functional properties had an 18% WSF component, 32 picks per inch, and pile ratio of 1:4.5.

Rao 1998, Gangopadhyay et al. (1999)

Direct warping is generally used for the preparation of ground warp beam while the use of the sectional warping machine for pile warp preparation is preferred. It is well-known fact that conversion process of yarn into fabric includes a number of intermediate operations, during which the yarn is subjected to varied amount of tensions and thus it is necessary that yarn meant for weaving must have adequate strength to bear with these tensile forces, failing which the process register low productivity.

Venkatpathi (1999)

Discussed that yarn quality in terms of strength alone is not adequate for the production of good quality fabrics. In addition to yarn strength, other factors which play a crucial role in deciding weaving efficiency are uniformity, the frequency of imperfections, long faults, slubs, count variation, twist variation, and hairiness. Further, he adds that many defects such as slack end, pulled warp, sticky ends, broken pattern, and so forth, predominantly originate in preparatory section.

Gangopadhyay et al. (1999)

Conducted study on assess manufacturing techniques of terry towels employed in the decentralized sector with a view to identify measures to enhance productivity and functional properties. They assessed machines, processes, and work methods followed at each stage of manufacturing such as winding, weft preparation, warping, sizing, weaving, and wet processing. The team suggested modification in machines and process to achieve desired performance in quality and productivity of terry fabrics.

Work by Terafdar et al. (2002)

Conducted studies for measurement of moisture transport in terry fabrics. In their study, they concluded that:

- Surface water absorption increases with the increase in pile height.
- With the increase in pile density, wicking behaviour increases both in warp and weft but quantitatively warp way show more wicking absorbency than weft way.
- With the increase in arial density of the fabrics, the wicking height increases.
- The wicking height also shows an increasing order with the increase in thickness of the fabrics.

5.4.3 Research studies on chemical processing of terry

Shenai (1981)

Grey terry towels contain various unwanted constituents such as contents of size recipe, natural wax and colouring matters, fragmented seed coats, leaves, stems, and so forth which need to be removed from the fabrics with a view to obtain desired functional as well as aesthetic properties in terms of softness, water absorbency, whiteness, and so forth. As all above-mentioned impurities are embedded in yarn, which is strongly interlaced in fabric, therefore, removal of these impurities from the fabric is a complex job. Treatment of fabrics with a suitable combination of chemicals at the required temperature and pressure for the specified duration helps in dissolving the deeply embedded impurities, while mechanical treatment in the form of stirring, rinsing, and so forth helps in separating out the dissolved impurities from the fabric/yarn. As the terry fabric contain piles/loops, therefore, the processing of these fabrics requires gentle treatment so that the configuration of piles is maintained in the desired manner.

Mahale et al. (1990)

He determined the quality of terry structure in terms of loop density, height, and the tightness of the weave. The absorbency of various terry fabrics is assessed in terms of sinking time, water uptake, and wick up methods. Except for the fabrics with loops only on one side, other terry weaves were found suitable for wear quality and absorbency.

Tendulkar et al. (1995)

They studied the prevailing practices of the bleaching process of terry towels and found that generally process flow given as under is followed:

i. Enzyme/acid treatment for de-sizing
ii. Washing
iii. Neutralization with acetic acid
iv. Bleaching in H_2O_2, caustic, and soda ash
v. Washing
vi. Softening
vii. Drying

Inputs by Gujar et al. (1992), Patel (1998), Gangopadhyay et al. (1999)

Kiers, winches, open tubes, and so forth, are conventionally used for de-sizing and bleaching for most type of fabrics. In this equipment, no specific arrangement was installed for treatment of delicate fabrics such as terry towels. Surveys conducted for the practices of chemical processing in the decentralized sector have revealed that still, raw methods are in use. But after the invention of 'soft flow machine', the effectiveness of chemical processing in terms of both cost and quality is increased a lot. In addition to soft flow machines, development of continuous dyeing range enhanced significantly the productivity and quality of terry fabrics and tubular knits.

Patel (1998)

He investigated the methods involved in terry towels' dyeing which includes such processes as side hemming grey fabric, bleaching, drying in open width, dyeing, washing, wet-on-wet finishing, weft straightening, and final drying, and batch-wise dyeing in rope form which involves scouring, bleaching, and dyeing in soft flow machine and final drying with pile lifting. Advantages of weft straightening during pad batch dyeing include no batch variations, little fabric shrinkage, and lower water and steam consumption. Advantages of soft flow system include lower investment costs, softer towels, lower dyeing cost, shorter process time, and low labour cost.

Teli et al. (2000, 2002)

They investigated the effect of different kinds of softeners and process of softening on functional properties of terry fabrics. Researchers evaluated four softeners with respect to fabric softness as well as other properties such as water uptake, sinking time, bending length, wicking height, whiteness, and hand. Researchers observed that 40°C was an optimum processing temperature. The terry soft – 478 softener exhibited the best overall performance.

5.4.4 Global leaders in terry towel production

The prominent loom manufacturers across globe include Vamatex, Saurer, Dornier, Nuore Pignone, and so forth. These looms were equipped with different aspects of mechanisms of pile formation such as use of microprocessor console, user friendly guide, touch screen facility and so forth has made to achieve controlled pick spacing, good cover, and so forth. Moreover, it is irony to note that the basic formation of the loop has remained the same despite the developments in loom design.

5.4.5 Indian perspective

This includes the spectrum of terry towel producers such as Bombay Dyeing, Modern Terry Towels, Abhishek Industries, Garware Wall Paper, Welspun Polyesters, Trimbak Industries, Sharda Terry Towel, Santogen Exports, Vanasthali Textiles, and so forth. In India, terry fabrics are manufactured mainly in decentralized handloom and power looms sectors (Gangopadhyay et al. 1999). Most of terry fabric centres are concentrated in south India such as Chennai, Erode, and Sholapur. Even though the organized sector has terry towel production, the percentage is less than that of the decentralized sector. Till last decade only 10–15% of total terry fabric production was produced in organized sector (Kwatra 1994). Most of the organized sector units are engaged in catering to market of export and high-quality segment of the domestic market, which covers mainly of hospitals, and leisure industry. Oversea manufacturers of terry include MARK Terry Ltd. Qualitex Group (BD) Ltd Jaantex Industries Ltd., Al-textile Mills Ltd., Bismillah Towels Ltd., Anaa Textile Ltd., Rajany Tex Ltd.

5.4.6 Approaches for production of terry pile fabrics

These fabrics can be produced either by weaving or by knitting, out of these two methods of terry fabric production, woven terry fabric, which is the first method invented, still has major share (Kienbaum 1978). This is because the quality of knitted terry fabric is not comparable to that of woven terry fabric. Besides the methods employed to manufacture the terry towels, other factors such as the use of fibres, parameters of yarn, parameters of weaving, and methods of chemical processing also play a significant role in determining the quality of terry towels (Swani et al. 1984, Teli et al. 2000). A terry or loop pile surface may be developed during the operation of weaving by each of three distinct methods namely:

a) Using terry pile motion, whereby during weaving several picks of weft are inserted a short distance from the fell of the cloth (or last pick inserted) to produce a short gap or fell, after which they are all pushed forward together to take their final place in the fabric. As each group of picks is thus pushed forward by the reed, pile warp threads buckle, or loop either on one side or on both sides of the cloth as per determined and so develop the characteristic loops of pile known as terry loop or uncut pile.

b) By means of wires that are inserted in the warp sheds at intervals (as if they were picks of weft) and subsequently withdrawn, thereby causing all warp threads that passed over them to form a corresponding no of loops.

Figure 5.5 Jacquard attached terry loom

c) By interweaving pile warp threads in such a manner that a looped or terry pile may be produced in an ordinary fast reed loom without employing a special terry pile motion and in which each successive pick of weft is beaten up to the fell of the cloth, as in the production of the fabric of ordinary construction.

5.5 Classification of towels

Towels are classified according to its weight, production process, pile density and pile presence on fabric surfaces, pile formation, pile structure and finishing. However, some basic concept of classification mentioned below the categories of end uses weave, and terry loop formation. Common terminologies used in the export market are mainly based on end uses. Commercially, three-pick terry towel is viable for all the aspects such as quality parameters, look, and appearance.

5.6 Terry looms

All the looms used for production of terry can be of different types namely handloom with dobby or jacquard, power loom with terry motion (either shuttle changing or cop changing looms – later is quite common as in former the efficiency is low due to mechanical problems), unconventional looms (air jet or rapier) with jacquard attachment. Various types of arrangements are shown in Figures 5.5 to 5.11.

5.6.1 Number of filling colours, warp stop, and types of weft insertion

Like in any other unconventional weaving, terry looms are provided with a weft colour selection system in which a maximum of 12 colours are used through a weft colour selection mechanism. Weft form the cone or cheese package is wound on to what are known as weft storage or weft accumulators prior to picking either by rapier or gripper or air jet media. The pick control mechanism or pick finder detects the weft breakage. At a filling break, the machine stops and moves at reverse slow motion – automatically – to free the broken pick. It has a significant role in reducing the down times for repairing filling breaks and thus the starting marks can be avoided (Picanol, 2004). Drop wires which are hung individually on each warp end, fall down when a warp end is broken or is very loose, closes down the electric circuit and thus shutting down the weaving machine (Baser, 2004).

Figure 5.6 Figured terry loom

Figure 5.7 Terry loom without JQ

5.6.2 Filling insertion with rapiers

Rapiers are popular in the production of terry cloth because of the flexibility they offer for production (Phillips, 2001). Rapiers are two hooks which carry the weft picks across the warp sheet. The first giver hook takes the weft pick from the yarn feeder and carries it to the centre of the warp width. Meanwhile,

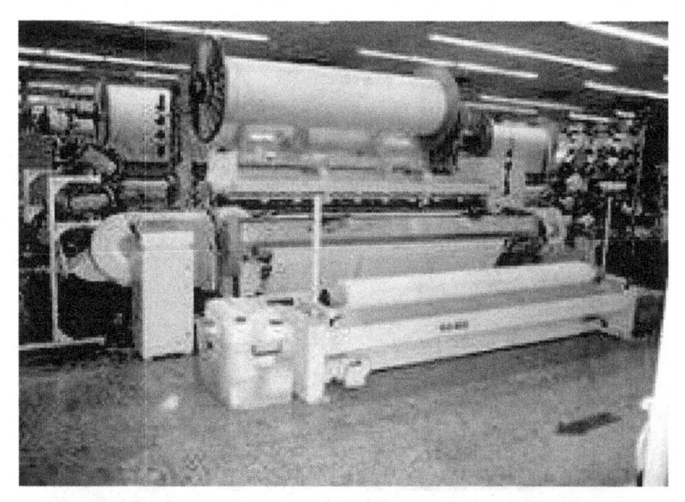

Figure 5.8 Terry loom without JQ

the taker hook moves from the other side of the weaving machine to the centre. There, the two hooks meet and the weft pick is transferred to the taker hook. After that, the giver hooks returns vacant to the side it came from, and the taker hook carries the weft to the opposite side (Adanur, 2001) In air jet, weaving a puff of compressed air carries the weft yarn across the warp sheet (Humpries, 2004). As seen in Figures 5.10 and 5.11, there are relay nozzles which are arranged in a definite order according to fabric width. These aide nozzles are connected to the main nozzles in groups. The air hoses which go to aide nozzles are also arranged in a row. The pick feeders also work with air and winds according to the fabric width. On the side where the pick

Figure 5.9 Weft Selecting process

arrives there are optical sensors which control the arrival of the filling picks. The maximum filling insertion rate practically achieved in terry weaving is 1800 m/min (Dornier, 2003).

5.6.3 Filling insertion with projectile

A small gripper takes the cut weft yarn across the weaving loom (Humphries, 2004). This system is not very common in terry weaving as rapier and air jet filling insertion system are most commonly used ones (Promatech, 2003; Dornier, 2003; Picanol, 2004; Smit Textil 2005; Tsudakoma, 2005).

5.7 Terry fabrics produced by means of terry motion

Any terry towel fabric has the following parts (if it is fully terry ornamented towel): fringes at the beginning and end, pile area, border, and selvedge. However, pile fabrics are also produced sometimes in hemmed fashion in which the ground and pile are woven with the same tension to get pile less headings. These are exemplified in so-called Turkish towels, bath mats, counter paved, toilet covers and mats, and many other articles employed for domestic purposes and in which the loops of pile occur with a more or less dense or close formation. Majority of the goods are produced entirely from cotton, although terry towels are sometimes produced entirely or in part from linen. Terry weaving is a principle imminently adapted to the production of towels, as the loops of pile give considerable bulk and import good absorption properties of the fabric.

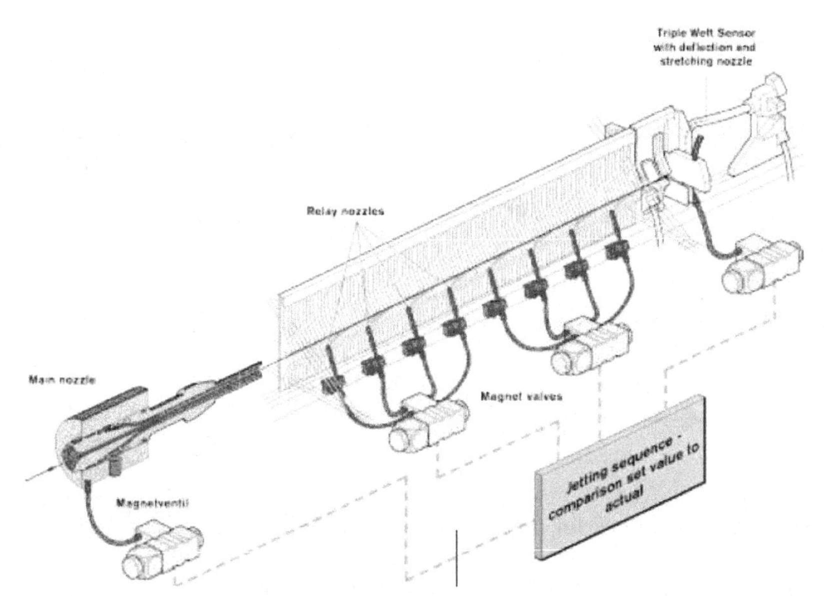

Figure 5.10 Air jet distribution system on terry air jet loom

5.7.1 Ornamentation of terry

As mentioned above, terry may be produced with loops on only one side – single-sided or double-sided terry in which we have both face and back piles. Again the fabrics may be either intermittent or interchanging type. Varieties such as stripe/check/figured are produced in either terry or Turkish fashion (single or double sided). When the pile is on the face only use of coloured pile threads enables stripes to be produced. When the pile is present on both sides then coloured pile may be used in such a way that face has one colour and back another. Simple checks to complex figures may be formed by causing the coloured piece threads to form loops first on one side and then on opposite side.

5.7.2 Loop or pile formation in terry and terry motions

Basically, the loop formed depend on the temporary gap formed between regular fell and false fell. A loop is formed when a thread is held at one end and pushed at the other end. Indeed, this is what happens in weaving (Fig. 5.11).

Terry weaving employs two warp beams namely ground and pile. Former is heavily tensioned and the later is lightly (except at pile less heading areas). The ground warp ends move forward slowly and under high tension as the ground warp beam turns slowly. At the same time, the pile warp ends move forward quickly and loosely as the pile warp beam turns faster than the ground warp

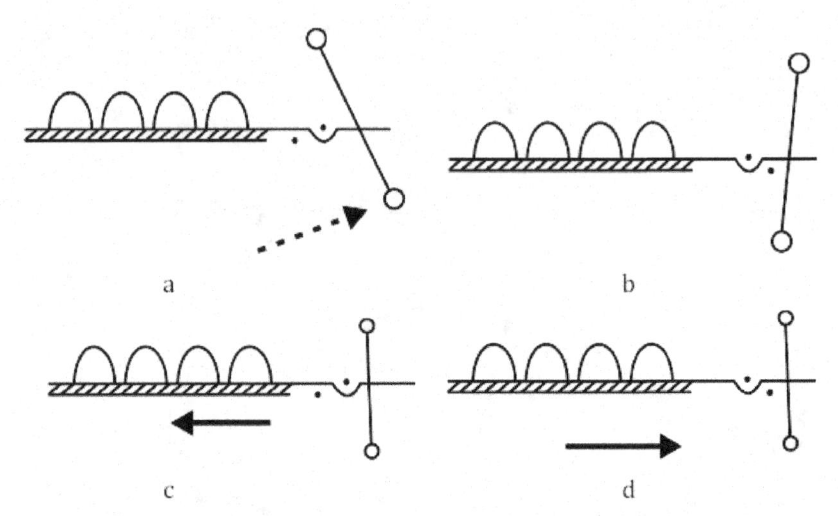

Figure 5.11 Stages for the formation of false fell in terry weaving

beam. Ground and pile warp beams are moved by two independent motors. Rpms of the pile warp beams is proportional to the required pile height. The higher speed delivers more yarn to increase the pile height. Ground end weave plain with the weft, whereas the pile end is slack which is held with one end in beam and the other end at fell. The free end is from weavers beam (pile beam) and fixed end is at fell. When reed slips or dragged back (As reed is made loose), no regular beat-up occurs and hence the weft inserted is allowed to stay or placed at a small distance from the regular fell known as 'false fell.' This is observed for the first and second picks. However, on the third pick all the three are beaten (wow the reed is fast) to fell of the cloth forcing the slack pile end to form loop (Figs. 5.11, 5.12, and 5.13). Terry may be produced with three, four, five, or six pick pile. As the number of picks changes the fast and loose picks also changes. Today majority of terry piles are three-pick pile type only. The object of inserting more number of picks for each row of loops is to produce a superior fabric and bind pile warp threads more securely to the foundation texture. During let-off, pile tension is controlled continuously. This decreases yarn breakages and avoids out-of-tolerance loop heights. In Figure 5.13a, the terry motion control system of Tsudakoma is shown. Here, pile tension is determined by pile tension roll which is propelled by a motor guided by electronic pile tension control system. The diameter of the pile beam is kept as large as the chassis of the machine allows so that it can hold the maximum length of pile warp. Keeping the pile beam's diameter large avoids changing the beam frequently during weaving. The width of the pile beam is between 76 and 144 inches (190–360 cm) and the diameter of its flange can be

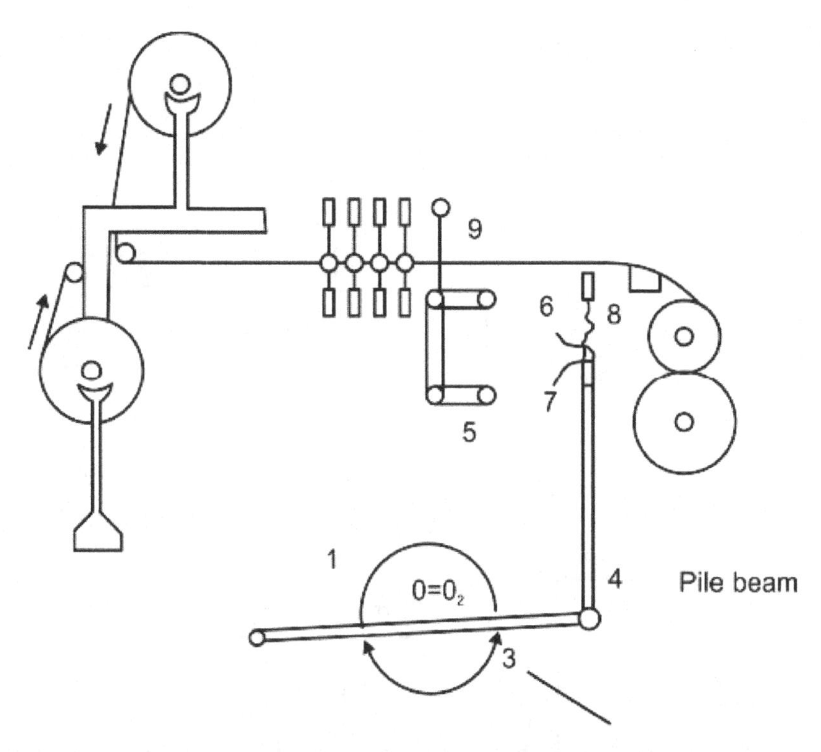

Figure 5.12 A typical terry motion

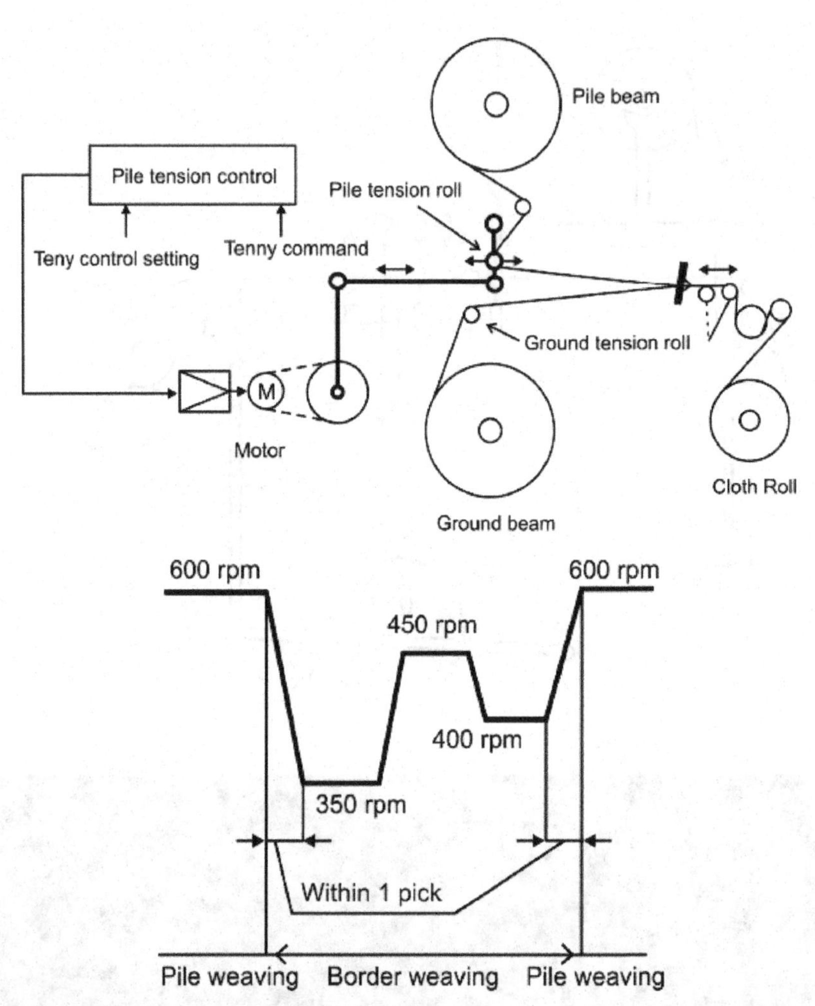

Figure 5.13 (a, b) Method of controlling pile yarn tension and speeds on Tsudokoma looms

up to 50 inches (125 cm), while the flange diameter of the ground beam is up to 40 inches (100 cm). The pile beam can hold more gross weight exceeding that of many automobiles (Rozelle, 1996; Dornier, 2003; Smit Textile, 2005).

The two warp systems are evenly let-off by a system of constant tension control from full to the empty beam. This is controlled by a highly sensitive electronic device. The tensions of the pile and ground warps are detected by force sensors, electronically regulated (Adanur, 2001). Elimination of unwanted increase of tension of warp during weaving high-density border and/or plain

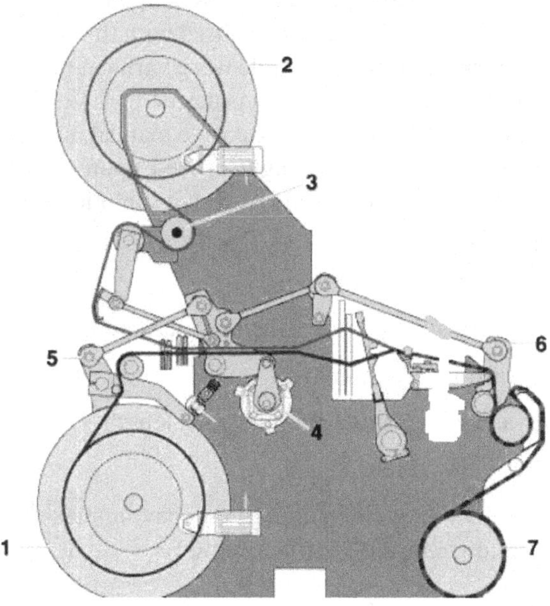

Figure 5.14 A modern type of Terry motion (a) Terry beams set-up, (b) set-up in Dornier terry loom (Courtesy: Dornier looms) 1 – Ground beam, 2 – Pile beam, 3 – Measuring unit, 4 – Terry motion unit, 5 – Sensitive backrest, 6 – Positively driven Emery set-up, 7 – Cloth roll (clutch controlled)

section is achieved by reducing let-off speed. In Figure 5.13, the diagram of pile warp tension in terry looms from Tsudakoma during weaving of pile, plain, and border areas are shown. In Figure 5.13b, the diagram of loom rpm's Terry looms of Tsudakoma during pile weaving and border weaving is shown (Tsudakoma, 2005). To prevent starting marks or pulling back of the pile loops, the pile warp tension can be reduced during machine standstill. An automatic increase in tension can be programmed for weaving borders to achieve more compact weave construction in order to ensure a rigid border and/or to achieve nice visual effects through jacquard or dobby designs on the border (Sulzer Rurti, 1998).

The way the backrest roller system is controlled depends on the weave. During insertion of the loose picks and during border or plain weaving, the warp tension between the open and closed shed is compensated by negative control. A warp tensioner with torsion bar is used for the ground warp, and a special tension compensating roll is used for the pile warp (Sulzer Ruti Inc, 1998). Precise measurement of pile warp consumption saves time, warp material, and thus money. At the same time, fabric quality is improved, which is highly competitive textile markets can only be advantage. Sultex has developed the new pile warp consumption monitor. Practically, any terry weaving machine can be equipped or retrofitted with this system. The device (patent applied for) measures the exact pile warp yarn length of a given repeat or cloth length in centimetres, with an accuracy of +/− 1%. Time-consuming, personnel intensive, and thus also cost-intensive manual measurement is thus no longer necessary, and the pile weight agreed upon with the client can nevertheless be adhered to exactly. Pile warp consumption monitor: mode of operation and displays

- Current reading
- Statistics of the six repeat lengths last measured
- Residual warp length in metres
- Set upper and lower limits for pile warp length
- Pick count within a repeat
- Red warning lamp and/or stoppage of weaving machine if the length last measured is outside specified limits

5.8 Take-up

The pick density is automatically controlled by synchronizing the take-up motor rotation with the loom speed. The take-up motor rotates the cloth pulling axle. The cloth pulling axle is covered with needles which pricks the terry fabric and assures that the thick fabric winds on the take-up roll evenly with a constant width (Acar, 2004). The electronically controlled cloth take-up guarantees exact weft densities in every terry towel and a faultless transition between pile and border (Smit Textile, 2005). There are five elements of a take-up system. These are:

1. **Temple**

   The temple holds the width of the fabric as it is woven in front of the reed and assures the fabric to be firm at full width (Picanol, 2004). A temple is seen in Figure 5.15 ensure the terry fabric is open to the sides and remains straight and tense throughout the fabric width.

2. **Length temple**

   Length temple is located on the centre of loom width between two side temples. There are groves starting from the centre and going to the left and right sides of the temple ensure the terry fabric is open to the sides and remains straight and tense throughout the fabric width.

3. **Cloth pulling axle with needles**

   It ensures the thick terry fabric keep its tension and width while being transferred from the length temple to the cloth transfer axle.

4. **Cloth transfer axle**

   It increases the contact angle between the terry fabric and cloth pulling axle with needles and transfers the fabric to take-up roll.

5. **Take-up roll**

   The fabric which comes from the transfer axle is wound on take-up roll (Acar, 2004).

5.9 Selvedges on terry loom

It is well-known fact that selvedges are necessary for any type of fabric and Terry is not an exception. Two types of selvedges are used namely Leno and tuck-in. Needless to mention here that terry towels are woven in three or four piece form depending on the width of each towel. Thus, selvedges and selvedge making units are very much necessary on these looms. Like in Sulzer projectile gripper loom of 330 cm RS, three or four shirting pieces are woven, terry looms also have similar feature as for as selvedge formation is concerned.

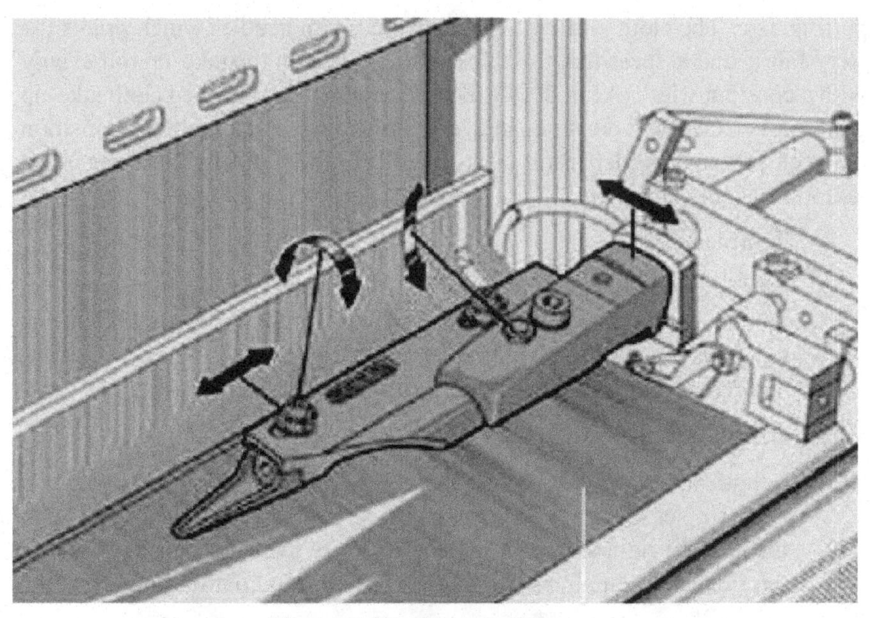

Figure 5.15 Temple for terry

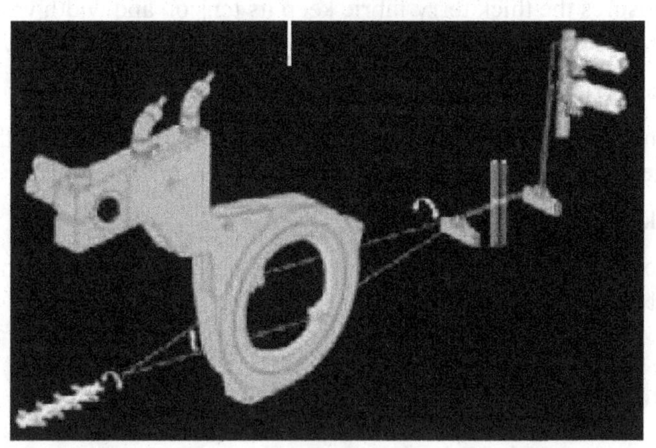

Figure 5.16 Formation of leno selvedge

5.9.1 Leno selvedge

A leno weave at the edges of the fabric locks in the warp yarns by twisting the last two warp yarns back and forth around each pick. They are made with special leno weaving harnesses (Adanur, 2001). Leno selvedges predominate in terry weaving (Dornier, 2003). In Figure 5.16, a leno selvedge forming system for terry weaving is shown.

5.9.2 Tuck-in selvedge

The fringed edges of the filling yarns are woven back into the body of the fabric using a special tuck-in device. As a result, the filling density is doubled in the selvedge area (Adanur, 2001). In Figure 5.17a, the needless tuck-in devices which are used in terry looms from Tsudakoma is shown. In Figure 5.17b, the diagram of the tuck-in selvedge is shown. Here, servo motor replaces the traditional terry cam for pile formation, so the reed does not drop back. When the reed is at the front centre, the fabric is positively driven towards the reed to form pile by the backrest and terry bar in combination with the temples (Seyam, 2004). The disadvantage of this system is that the friction which takes place during the forward–backward motion of the ends can lead to end breakage. Although weaving machines of different makes have different mechanism the main principle is the same (Adanur, 2001).

5.10 Types of terry motions (fast reed or loose reed)

The variable beat-up motions are an essential part of the terry pile weaving and they fall into two main categories. The function of these motions is mentioned earlier to create a gap between the cloth fell and the first two picks of the pile forming a group of picks termed loose picks as opposed to picks beaten up fully which are known as fast picks. In the first category are those mechanisms in which the reed itself is drawn back on the loose picks these by leaving them a small distance short of the cloth fell. A variety of device exist to achieve this purpose in some of which the reed only and in others the slay itself may be controlled to provide a short beat-up. On the following pick, the reed or slay is locked fast so that the preceding loose pick and the fast pick are pushed together into the cloth fell proper. The two reed positions are shown in Figure 5.11. The arrangement as shown in Figure 5.12 illustrates one type of terry motion which allows the reed to go loose and stand firm alternatively. The plate wheel 1 carrying the bowl 2 acts upon a level 3, every third pick and pulls down the rod 4, and the bowl 5, passing between 6 and 7 holds the reed firm. On the two intermediate picks, when 2 is not acting, the spring 8 lifts 4, and the bowl 5 not being able to enter between 6 and 7 the reed 9 is forced out of position. (Fig. 5.14b show Dornier set-up) In the second category of mechanism, such as are used present in gripper and rapier mechanism, the reed is permanently fixed in position and has an instant stroke. To create the gap on the loose picks, the cloth itself is drawn away from the advancing reed so that the two loose picks cannot reach the normal cloth fell position. On the third pick, the cloth is brought forward again so that the three picks of a group join together with the previously woven cloth at the normal cloth

Figure 5.17 (a, b) A special type of tuck-in on Tsudokoma looms

fell. Here, servo motor replaces the traditional terry cam for pile formation, so the reed does not drop back. When the reed is at the front centre the fabric is positively driven towards the reed to form pile by the backrest and terry bar in combination with the temples (Seyam, 2004). The disadvantage of this system is that the friction which takes place during the forward–backward motion of the ends can lead to end breakage. Although weaving machines of different makes have different mechanism the main principle is the same (Adanur, 2001).

5.11 Terry designing

Terry fabrics are often very complex with different coloured warp ends in combination with loop patterns. They are subject to changing fashions, and the market is constantly demanding new qualities and designs. The rapid development of electronics has enabled fabric designers to produce completely different patterns. Through a servo motor, the beat-up position for each pick, and, thus the type of terry and the pile height can be freely programmed from one pick group to another. In this way, nearly 200 different loose pick distances, and hence the same number of pile heights, can be programmed

Figure 5.18 A terry pattern achieved by CAD (Computer Aided Designing)

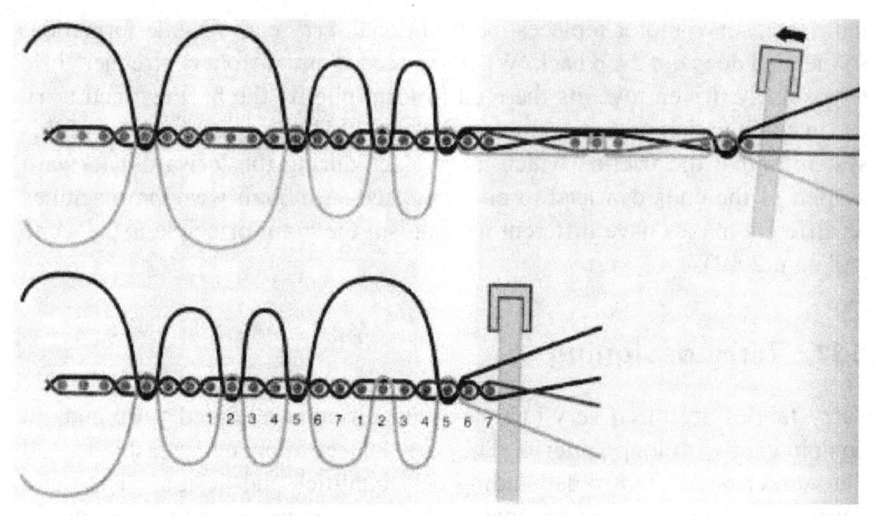

Figure 5.19 Cross section of new designed terry weaving two different heights of loops (Smit Textile, 2005)

In this way, a second pile height is also formed in filling direction, making sculptured patterning possible by the difference in pile height in the warp, and filling direction (Adanur, 2001). A requirement for this kind of pattern formation is a freely programmable sley traveling on a rapier weaving machine. Microprocessor control allows the loose pick distance to be programmed easily and individually for each pick (Sulzer Ruti, 1998). The loop formation system with full electronic control lets you alter the height of the loop by accompanying the electronic weft ratio variator device on jacquard looms to program different weft ratios like three- pick terry, four-pick terry, and so. By this method, different heights of loops can be achieved in the same shed (Promatech, 2003).

5.12 Classification of terry towels

Basically, terry towels are classed as power loom woven and handloom woven. In turn, these may be either plane terry (without any designs), intermittent terry, interchangeable terry, or figured terry. Terry towels are classified according (Gungor et.al.) to weight, production style, finishing, number of picks in a pile formation, and pile presence of the fabric surfaces. Based on weight, they are identified as more than 550 g/m^2 as heavy weight, 450–550 g/m^2 as heavyweight, 350–450 g/m^2 as medium, 250–350 g/m^2 as lightweight terry towels, respectively. Depending on the method of fabric production they are identified as woven, weft knitted, or warp knitted terry towels. Towels are also classified based on finishing as velour, printed towel with embroidery, and towel with appliqués. A towel with appliqués is embellished with additional pieces of decorative fabric in a motif which is stitched onto the towel. In velour towel, pile loops on one side of the fabric are sheared in order to give a smooth cut velvet appearance. Uncut loops give the best absorbency, whereas velour gives a luxurious velvety hand. Two-pick terry towels which are woven specially for bathrobe end use have lost their importance due to the instability of loops and hence today most of the terry towels are of three-pick type only. Depending on the number of pick per one terry cycle they are identified as Three-pick plie, four-pick, five-pick pile, and six-pick pile, respectively. Plies may be arranged only one side giving 'single-sided terry' and on double side giving 'double-sided terry', piles may be observed on face for some time and then at back giving 'Intermittent terry', interchanging face and back at regular intervals giving 'Interchanging terry' and so forth. One-sided pile is not preferred because it has low water absorbing capacity and is only meant for some special application and further weaving process without defects is an uphill task in one-sided case. Towels are also classified as follows

Groups such as bath towels, face towels, fingertip towels, kitchen towels, and washcloths based on end use and size.

Or

Terry is classified into various basis and one such basis is as follows

Depending on use: Bath sheet, bath towel, hand towel, wash towel, face towel, guest towel, bath matt, bath robe, kitchen towel.

Depending on terry loop configuration: two, three, four, five, six, and seven pick terry.

5.13 Fibres suitable for terry towels

Generally, terry towels are used as bathing towels where its main function is to absorb water during the course of wiping of the wet body. It is therefore fibres that have greater ability to absorb water as well as have softer feel are preferably used in the manufacturing of terry towels. The two fibres that have quality matching with the requirements of towels are cotton and linen. Linen has a rather harsher feel but it may not be a disadvantage in certain cases. The price of linen is also a barrier. Viscose staple yarn is also used as it possesses adequate moisture absorption capacity but its ability to resist frequent laundering is poor as compared to cotton. It is, therefore, the bulk of towels are manufactured from cotton.

5.14 Basic parameters of a quality terry towel

Weight and GSM: weight and GSM (Grams per Square Meter) should be the same as required by the customer. Every manufacturer has some template or software (ERPs) where towel manufacturers calculate everything likes pile's height, the density of picks, and ends to meet the requirement. This database or any software has been developed through some basic calculation.

Softness/Hand feel: it depends on properties of the yarn used in pile, finishing chemicals, and to some extent on pile orientation.

Pile orientation: totally depends on the process line.

Lint: lint are basically protruding fibres present in a finished towel. It is measured by the weight of accumulated fibre collected from the washing machine and tumble drying machine during testing.

Absorbency: terry towel should be highly water absorbent.

Dimensional stability: how a towel is behaving after washing is fall under dimensional stability properties. Dimensional stability is measured by the residual shrinkage % in a finished towel.

Other parameters are strength, color fastness, and so forth.

5.15 Designing of terry towel and the requirements of a terry towel

A terry towel should exhibit high absorbency, high wet strength, ability to dye with various colours having affinity with fibre structure, good colour fastness, good wash ability, soft hand, hypoallergenic, low cost, and easy production and availability. For towelling 0.93–1.1 inches staple length with 20 32 g/tex fibre strength, 3.5–4.9 micronaire and 0.8–0.9 maturity ratio cotton fibres are suited best.

Fibres other than cotton

Today terry towels are being produced from different fibres (100 %) such as modal, mamboo, seaweed, lyocell, soyabean, corn or blend of bamboo, silk and cotton, and so forth. Bamboo is preferred next to cotton because of its softness, high absorbency, luster, antibacterial properties. It is not exaggerated to mention here that even terry towels made in microfibres are very popular as they possess ultratouch/high absorbency made from synthetic fibres in blends such as polyester and nylon. These products will absorb water by five to seven times their weight in water. Towels made from microfibres are available in different colour and weave, and so forth. The heavier the microfibre the more water it can absorb. Compared to ringspun cotton, microfibre show higher absorbency. In a recent study by Editor Murphi (2006), the potential of twistless yarns in terry towel production is reported and results of the study show terry towels are satisfactorily produced using twistless yarns. Very recently organic cotton is used for terry production.

5.16 Type of yarns in terry

Yarns used in terry weaving may be grouped as grey yarns and processed yarns (bleached, mercerized, or dyed yarn).

5.16.1 Pile yarn

Generally, a terry towel uses double yarn in the warp and a single yarn in the weft. The pile yarns used may be either 10's, 14's, 16's, or 20's singles or 2/20's, or 2/24's or 10's + 60's PVA. Yarn made of PVA can be doubled with cotton yarns for the production of soft terry fabrics. PVA fibres will be dissolved out during processing stage, leaving a fluffy soft terry structure with soft twist. Use of double or folded yarn will increase absorbency and fabric show resistance to pile lay. The use of two-ply yarns improves the visual appearance. Plied yarns are used to form upright loops in classic terry, whereas single yarns are used to form spiral loops in fashion terry known as milled or filled goods. Classic terry patterns are usually produced by using dyed yarns (package dyeing) and fashion towels are piece dyed or printed. Rotor yarns are also used as pile warp as these are known for high uniformity and absorption.

5.16.2 Ground yarn

10's singles or 2/20's or 2/24's English count yarns are used as ground with 100 % cotton. It is preferable to use double yarn as ground because it is subjected to high state of tension during weaving.

5.16.3 Weft yarn

12's or 14's singles or 16's singles or 20's singles or 2/20's or 2/24's or 2/16's yarns are used as weft. Normally, rotor yarns are weft yarns in terry due to their high evenness. However, it is necessary to make some arrangements for controlling the weft breaks due to poor strength of rotor yarn as compared to ring yarn.

5.16.4 Border weft

Premium or high-quality terry use decorative, shiny, and bulky yarns made from rayon or polyester or mercerized cotton in different counts. The borders are complex with fancy weaves. A model for pile yarn quality is mentioned below. However, deviations from the table are made depending on the end use or market requirement. Count CV % less than 1.2, strength CV % less than 5, CSP nearly 2400–2500, soft twist, Rkm 14–15, elongation % 4–5, elongation CV % less than 10, imperfection/km 20–40, hairiness index less than 10.

5.17 Parameters of yarn

It's necessary to consider the parameters of yarn irrespective of fibres as these controls the performance of the terry towel directly. Among the various parameters of yarn, linear density, twist, and packing density play a significant role in achieving the desired quality of terry towels. Higher packing density lesser will be the absorption as lesser space is available for water retention.

5.18 Formation of pile

Terry pile or Turkish pile fabrics are characterized by the formation of temporary fell or false beat-up. The formation of the pile is explained in the following texts.

The pile height depends on the number of picks inserted as loose and fast. In other words, the pile height depends on the distance of temporary or false fell formed from the regular fell.

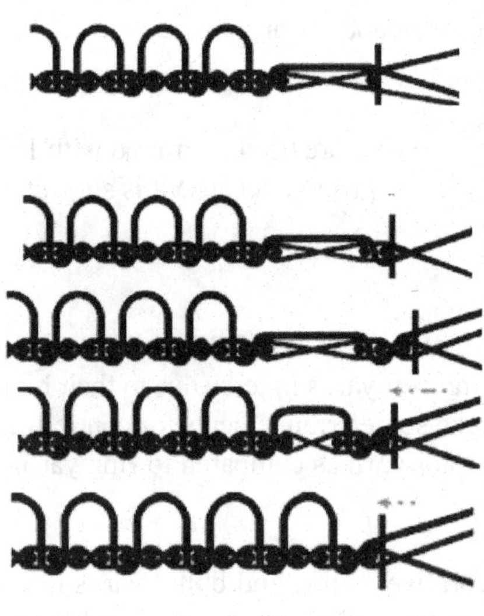

First pick is inserted and it will not be beaten to fell of the cloth or it is known as 'loose pick' and the reed will be swinging backwards in loose reed motion or the cloth is taken forward

Insertion of second pick and still the temporary fell unmoved

Third pick inserted and Reed is about beat all the three picks

Slowly the entire group of three picks will be pushed towards fell

Full beat-up of all the picks. Formation of regular fell as all the three picks inserted are beaten to the fell of the cloth due to the fast position of Reed

Figure 5.20 Stages of formation of pile

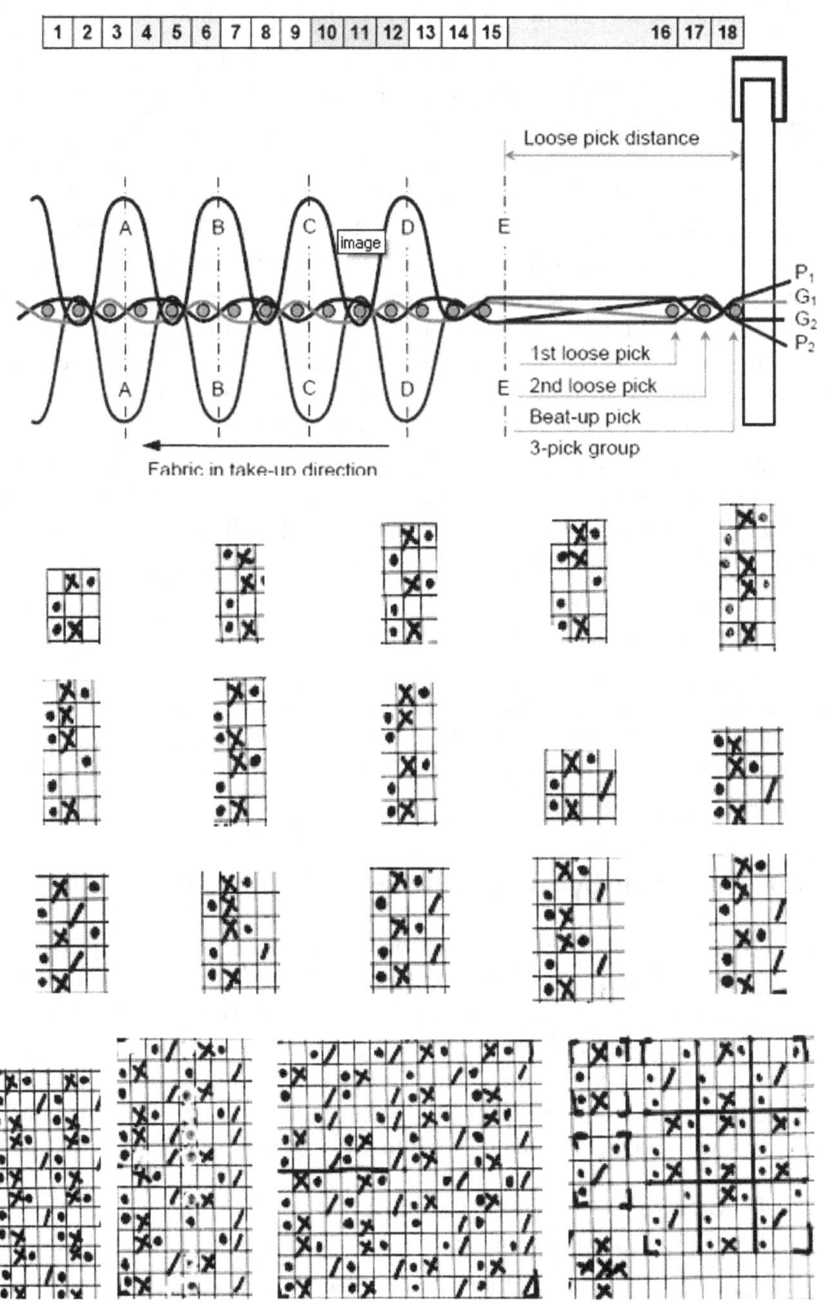

Figure 5.21 Cross section of a 3-pick pile

The complete formation of the loop is explained as follows.

The exact relation of the weft to the two warps and the principle of loop formation are depicted by means of the weft section in Figure 5.20. The broken vertical lines CC, DD, and EE divide the first, second, and third picks into repeating groups of three, line EE indicating the position of the fell of the fabric. On the right of the diagram, a group of three picks, which compose a repeat, is represented previous to being beaten up to the fell of the fabric. The ground threads G1, G2, and the face and back pile threads P1 and P2 are shown connected by lines with the respective spaces in the corresponding weave given in Figure 5.21. In weaving, the fabric, the group warp beam carrying the threads G1 and G2, is heavily tensioned. As stated earlier these threads are held tight all the time. The picks 16 and 17 are first woven into the proper sheds, but are not beaten up fully to the fell of the fabric at the time of insertion in their sheds; but when the pick No. 18 is inserted, the mechanisms are so operated that the three picks are driven together into the fabric at the fell EE. During the beating up of the third pick the pilewarp threads P1 and P2 are either given in slack or are placed under very slight tension.

5.19 Figured terry

In this class of terry, the loops follow a picture or a motif and the terry may be single or double sided. Figured terry may also be either interchanging or intermittent type. The majority are the former type. Value addition may be improved using coloured threads.

5.20 Technology of terry towel production

Terry towels are produced both in organized and unorganized sectors. However, the terry towels woven on handloom can be very well identified by the poor weft density or beat-up. But most of the towels in southern India are being produced on handlooms. It is well-known fact that handloom woven fabrics exhibit poor fabric cover and it is 100% true in terry towels also. The figured terry is normally produced on power loom jacquard. The technology of terry towel production involves the following main stages.

5.20.1 Spinning

Regular cotton spinning with blow room, carding, draw frame, simplex and ring frame, or open-end spinning is used. In the later, rotor spinning or twistless spinning is used to produce the warp and weft yarns meant for terry towel production.

5.20.2 Preparatory process to weaving

It includes the coning on machines such as Autoconer for quality packages so as to get maximum production efficiency at warping, use of high-speed beam warper with V-creel or Benninger creel with quality warp beams so as to result in zero lappers at sizing. Depending on the pattern of warp, colour packages may be arranged in warping creel.

In sizing use of multicylinder sizing machine with pressure cooker, storage kettles, and double sow box is recommended. The size mixture consists of ingredients such as maize (25 kg), anilose (25 kg) (a soluble starch or modified starch or thin boiling starch), PVA (9 kg), CMC (4 kg), gum (2 kg), Sio texcise (2 kg), mutton tallow (2.5%*), urea (2%*), French chalk (4 kg), and so forth, to get nearly 12–14% add-on for single yarns and for double yarns 3–4% add-on is planned. Details of sizing recipes and process, reader is directed to refer Acar, 2004. In postsizing operations use of fully automatic drawing – in machines and denting – in machines is recommended. Use of all metal reeds with ground and pile ends in the same dent based on the RTP is recommended. Open type of drop wires are mounted once the beam is gaited on the loom. Terry weaving is practiced on both conventional and unconventional looms. If it is conventional power looms, loose reed mechanism is fit so that loose and fast picks are inserted depending on the number of loose and fast picks. Majority of terry towels are produced on rapier and air jet looms today in India.

5.20.3 Weaving arrangements

Terry towel production includes the formation of false fell as explained above and during this, the reed may be either loose or fast. In the former case, it is referred as 'loose reed' motion and is found on all power loom. All the unconventional looms are developed with 'fast reed motion.' But among all the motions in weaving, it is let off motion which plays a significant role in terry production either on the conventional or unconventional loom.

5.20.4 Let off (conventional looms)

Terry production involves two types of areas known as pile heading and pile less heading. In pile heading, pile thread is slack and fast delivery is ensued during formation of the pile or during let off the extra length of yarn is let off depending on the length of the pile required. But in pile less heading both ground and pile weave plain or ground weave together and beat-up is on all picks. The control of the tension of the pile beam is effected through cross-border dobby.

---

* = as a percentage of Starch

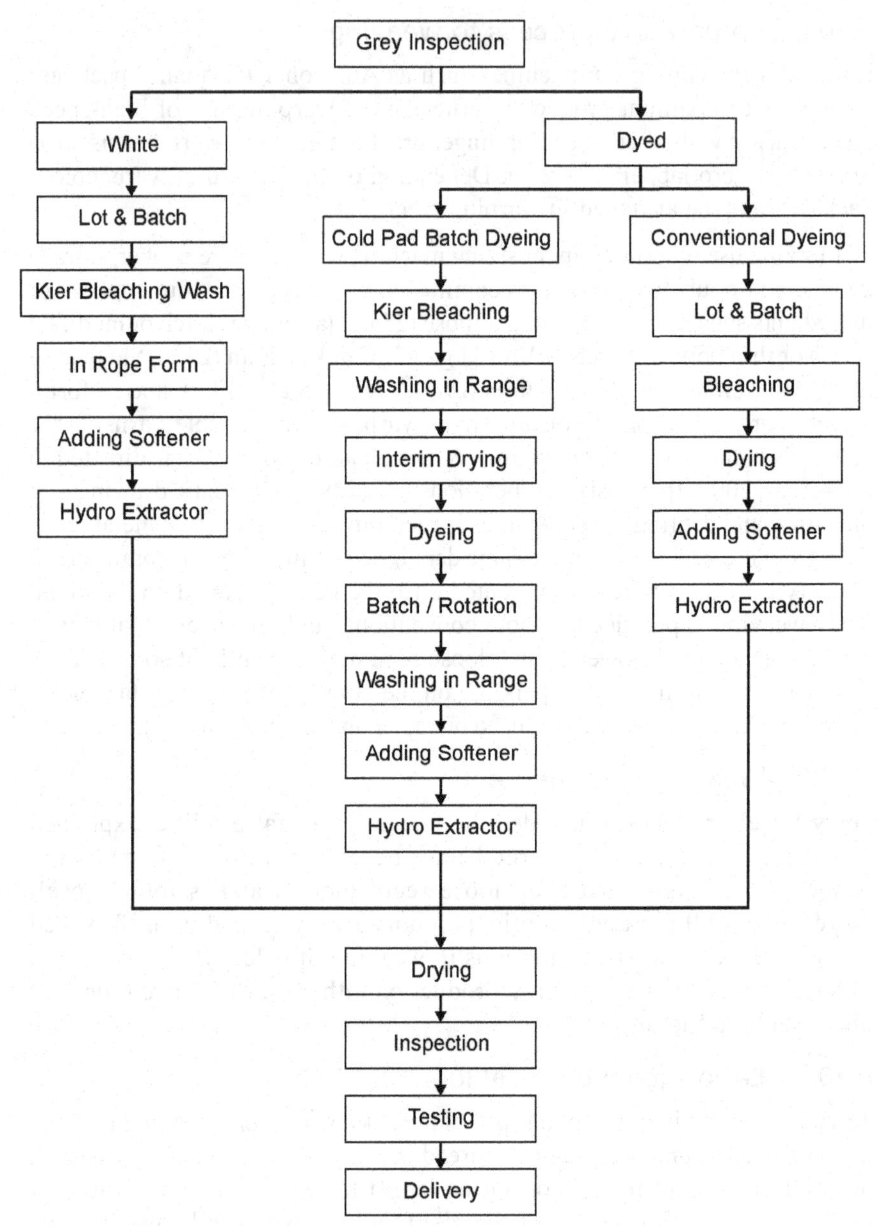

Flowchart 5.1 Wet processing flow chart of terry towel

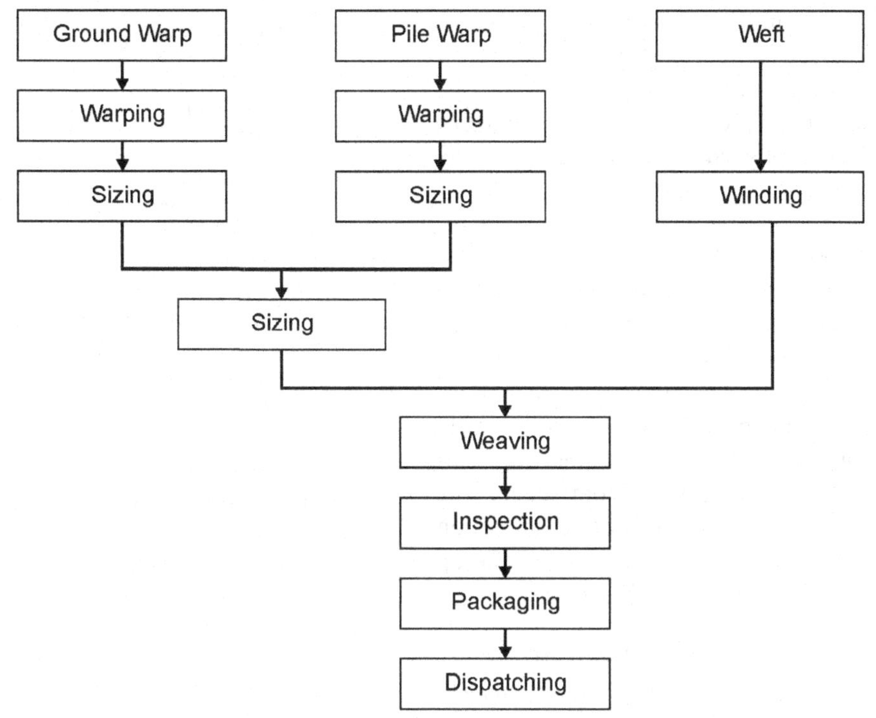

Flowchart 5.2 Showing the flow chart of terry production

5.20.5 Let off (unconventional looms)

All the modern looms are fit with a unique terry motion which includes the control of the tension of pile beam as and when required. Ground warp threads move slowly as they are under high tension. Pile warp ends move fast as they are under slack condition. Ground and pile warp beams are moved by two different independent motors. The speed of pile warp beam is proportional to the required pile height. The higher speed delivers more yarn to increase the pile height. In a typical terry motion found on modern looms, the pile tension is determined by pile tension roll which is propelled by a motor guided by electronic pile tension control system. The diameter of the pile beam is kept as large as possible to minimize the beam replenishments. The width of pile beam is 190–360 cm or 76–144 inches and the diameter of the flanges may be upto 50 inches. The two warp systems are controlled evenly from D_{max} to D_o. Sensitive backrest control the rate of delivery of warp depending on the warp tension during weaving. The tension of ground and pile is sensed by force sensors and thus the process is controlled thereof. To prevent the thin places or starting marks, it is advisable to reduce the pile warp tension during machine stoppage. In Sulzer Ruti machines, the sensitive backrest mechanism working depends on the weave employed in terry. The warp tension of ground and pile is separately controlled by sensors.

5.20.6 Take-up

The pick density in terry weaving is very low and some extra devices such as cramming are employed to increase the density of picks. In modern looms, the pick density is automatically controlled by synchronizing the take-up motor speed with loom speed. The electronically controlled take up motion ensures constant pick density. Take up motion uses full-width temples.

5.20.7 Selvedge motion

Either tuck-in selvedge or leno selvedge motion is found on terry weaving and depending on the sizes of the towels the selvedge motions are arranged thereof.

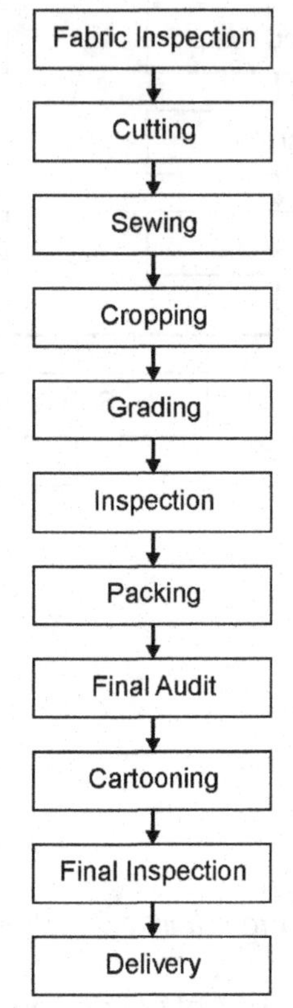

Flowchart 5.3 Stitching flow chart of terry towel

5.20.8 Weft measuring and patterning motion

In all the modern looms weft accumulators are provided and are selected based on the weft pattern. About 12 colours are being used today in terry weaving.

5.20.9 Warp and Weft thread control

In the event of break pick finder detects the weft breakage and adjusts the cloth fell accordingly reducing the chances for starting marks. Electrical types of drop wires stop the loom in the event of warp breakage.

5.20.10 Gery fabric inspection and rolling

All the grey fabrics are subjected to inspection if online fabric inspection is not available and rolling is carried out so as to prepare for chemical processing of the grey terry towel fabric.

5.20.11 Post-weaving of terry towels

Shearing

It is quite common practice to shear the terry loops after manufacture in order to create a cut-pile effect. Many hand towels are sold with one face showing the traditional terry loop, whilst the other side shorn to give the velour effect (Donaldson, 2003).

Shearing is applied to the pile fabric, by passing it over a cylinder with blades such as a giant cylindrical lawnmower. The velour fabric is then brushed with bristles set in a cylinder to remove cut bits of fibre. Brushing leaves the surface fibre lying in one direction so care must be taken to have all the fabrics in the same batch laid out in the same direction, or light will reflect off various pieces differently (Humpries, 2004). Shearing is applied to the pile fabric, by passing it over a cylinder with blades such as a giant cylindrical lawnmower. The velour fabric is then brushed with bristles set in a cylinder to remove cut bits of fiber. Brushing leaves the surface fibre lying in one direction so care must be taken to have all the fabrics in the same batch laid out in the same direction, or light will reflect off various pieces differently (Humpries, 2004). In Figure 5.22, a simplified diagram of the shearing process is given. The pile fabric is guided across the shearing table and is sheared between the shearing blades mounted on a cylinder and a fixed blade (Rouette, 2001).

Sculptured or carved design

The sculptured design is different from the one which is achieved during weaving by using long and short loops. This involves considerably processing after weaving. The pile fabric which has been woven with single pile loop height is first embossed, then the pile left upstanding is sheared off, and that

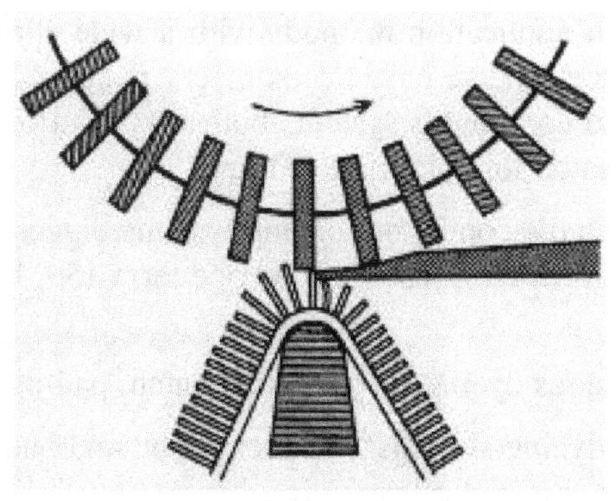

Figure 5.22 A diagram of shearing process (Rouette, 2001)

which was flattened is brushed up, leaving the sculptured, or carved design (Humpries, 2004).

5.21 Chemical processing of terry towels

The grey terry towel is subjected to desizing, scouring, and bleaching prior to dyeing.

5.21.1 Desizing

Enzymatic desizing

Starch is removed by the enzyme. This desizing process simply involves liquefying the film of size on the product. Bacterial, malt, and pancreas amylases are used as desizing agents. Enzymatic desizing is the classical desizing process of degrading starch size on cotton fabrics using enzymes. Enzymes are complex organic, soluble biocatalysts, formed by living organisms that catalyze chemical reaction in biological processes. Enzymes are quite specific in their action on a particular substance. A small quantity of enzyme is able to decompose a large quantity of the substance it acts upon. Enzymes are usually named by the kind of substance degraded in the reaction it catalyzes. Effective enzymatic desizing requires strict control of pH, temperature, water hardness, electrolyte addition, and choice of surfactant. Bacterial enzymes are preferred because of their activity over a wider pH range and tolerance to variations in pH. Since desizing is carried out on grey fabric, which is essentially non-absorbent, a wetting and penetrating agent is incorporated into the desizing liquor.

5.21.2 Fabric dyeing

Reactive dyes are preferred due to bright shades, good fastness properties, easy application, moderate cost, and eco-friendliness Reactive dyes offer a great flexibility in application methods with a wide choice of equipment and process sequences and so have become very popular. These are applied through exhaust and continuous systems both very comfortably. Any of the following are suggested for colouring of terry.

a) Exhaust/batch/dis-continuous dyeing systems: Jigger open width 3–5:1, Winch rope form 20:1, jet/soft flow rope form 15:1, beam dyeing open width 10:1,

b) Semi-continuous dyeing systems: pad-batch, pad-jig and pad – roll.

c) Continuous dyeing systems: pad-thermosol, and pad-steam

Reactive dyes are cheaper as compared to other class of dyes.

5.21.3 Printing

Various printing types such as direct printing, discharge printing, and resist printing and techniques such as roller printing and full-screen printing are available for terry printing.

5.21.4 Final finishing of terry towels

Final finishing includes all the finishing treatments applied to the fabric after dyeing and printing it can be divided into two: chemical (or wet) treatments and mechanical (or decorative), treatments. Chemical treatments include softening, hydrophilic, and antimicrobial treatments are among the chemical finishing of terry towels.

5.21.5 Hydrophilic treatment

Silicones are added to the towel to give hydrophilic properties and soft handle. It is also used to give a soft handle.

Softening

The three basic types of softeners which are used on towels are cationic softeners, non-ionic softeners, and silicones. Cationic softeners give good softness, are only used for coloured towels. Non-ionic softeners have less softening effect but are used in white towels. Silicones are the best and the most expensive of the softeners Hydrophilic silicones also affect the hydrophilicity of the towel positively. There are also applications of enzymatic softening using cellulases.

5.21.6 Antimicrobial treatment

Towels can be treated with antimicrobial finishes in order to prevent mildew attack, reduce odour, and minimize the spread of harmful organisms Two types of antibacterial and deodorant finishes are available The first is applied during fibre-forming process, whereas the other is incorporated into the finishing process. The second approach is more versatile and widely adapted. Chemical entities are responsible for imparting antibacterial attributes including fungicides and bactericides. Obtaining antimicrobial properties by using antimicrobial fibres is achieved by anchoring the antimicrobial agent in the fibre. Trevira bioactive (R) is an example of antimicrobial fibre used in towels which have proven to fully retain its antimicrobial effect after 100 domestic or 50 commercial wash cycles.

5.21.7 Mechanical

The main aims of dry treatments are to give the towels fuller volume, and dimensional stability and dryness.

Tumble drying

The towel is given a fluffy and soft hand, and some particles are removed during drying. The common way is to use continuous tumbler dryer. The second way is to use tumble dryers which are a huge version of domestic tumble dryers.

Stentering

Stentering or tentering is a controlled straightening and stretching process of cloth to improve the dimensional stability of either pin or clip or hook stenters are used. During the process, the cloth is slowly and permanently brought to desired condition.

Cutting and sewing

In this stage, the terry towels pass through four steps such as longitudinal cutting and longitudinal hemming.

Cross-cutting and cross hemming

These processes are achieved by scissors and standard sewing machines. Lengthwise cutting machines are used for the first step of this stage. Longitudinal cutting of towels which have been produced on the loom as several panels joined side by side. In these machines, there are several cutters which cut lengthwise between adjacent towel panels in order to separate them. Longitudinal hemming is achieved by lengthwise hemming machines most of which are usually equipped with two 401 chain stitch sewing machines. Labels can be attached during this process. Towels then pass through cross-cutting as the third step. Transversal cutting machines carry out product stacking and automatic discharge. The cut product is stacked in layers one on the other. This is followed by cross hemming. In both length and crosswise hemming, the hems are first folded and then sewn. In the lengthwise hemming, the border constitutes a problem due to its higher thickness. Pile sewing is avoided by the pile raising device. Labels can also be attached in cross wise hemming. Some of the hemming machines use 301 lock stitch instead of 401 chain stitch. Overedge stitching is also preferred to conventional hand hemming. In automatic lines both cross-cutting and cross hemming are carried out on the same machine. Various stages in final terry preparation is as follows

5.22 Quality control aspects of terry towels

Terry weaving can include two methods of inspection such as online and offline. Online inspection is aimed to reduce the defects during the fabric manufacture itself. Offline is the conventional practice and backward integration is required to control the quality of terry towels. Here, two aspects need to be considered. One related to process control leading to good quality and other quality control to ensure quality. As far as process control is concerned the same old points of cotton processing on winding, warping, and sizing machines holds good as the count employed for terry is the routine one. However, following defects are identified in terry towel weaving. Care should be taken to produce a quality towel.

Weaving defects: missing pile, missing pick, thin or thick pick, beginning or end part missing, side part missing, stop mark, reed mark, wavy or curly selvedge, dense pick spacing, and so forth.

Wet processing defects: uneven dyeing, shade variation, off shade, print defect, mismatch in print, design defect, noxious odours, unfinished chemicals, and so forth.

Sewing defects: low stitch number, widths of parts out of specification, missing stitch, overlapping stitching threads.

General defects: irregular dimension, oil stains, colour stains, holes, cuts, crack, jelly, crushed piles weight variation, pile height variation, and so forth.

Control of weight in terry: As terry fabrics are sold on weight basis, it is necessary to control weight per unit area during weaving. The range for terry weight is 250–700 g/m^2 and generally terry manufacturing use standard yarns in warp and weft. Hence to achieve the required weight per unit area picks per inch and pile length are significant parameters to be considered. The length of the pile in relation to the length of terry fabric is measured in terms of pile ratio. Following expression is useful to understand terry geometry

Pile height (cm) = (Pile ratio/Picks per cm) × 0.5 × type of terry

5.23 Economics of terry production

Production of loom (towel/per day/mc) = Rpm × 60 × 24 × efficiency/picks per towel

Pile height = pile ratio × 15/picks per cm (for 3 pick terry towel-commercially it is running)

Pile ratio = (pile weight in gram × 2.2046 × 840 × pile count × 36)/(pile ends per towel × length of towel × 1000)

Finished weight of Towel = (Gsm × towel length × width in cm)/10000

Weight per dozen (Lbs/doz) = (finished towel weight × 12)/453.6

Let total number of ground warp ends be 694,
let ground warp count be 25 tex × 2

Let warp crimp be 8%, weft yarn count = 34 tex,
number of pile warp ends = 576

Length of pile part = 102 cm, pile ratio (for pile height)= 52:10 (52 cm of pile warp for 10 cm of cloth), pile yarn count = 30 tex × 2, length of plain part = 4 cm, picks per cm = 20

Reed width = 58.4 cm, grey length (pile and plain part) = 106 cm, fringe length = 2 cm

Then, weight of ground warp

$= (106 \times 1.08 \times 694 \times 25 \times 2)/(100 \times 1000), = 397.25$ g

Weight of ground warp in fringe (It is assumed here the warp crimp role is zero)

$= (2 \times 694 \times 25 \times 2)/(100 \times 1000) = 0.69$ g

So total ground warp weight = 397.25 +.69=397.94 g.

Weight of pile warp = weight of pile warp in pile part + that in plain part + that infringe

a) Weight of pile warp in pile part (pile ratio: 52:10)
$= (102 \times 576 \times 52 \times 30 \times 2)/(100 \times 1000) = 183.31$ g

b) Weight of pile warp in plain part
$= (4 \times 576 \times 1.08 \times 30 \times 2)/(100 \times 1000) = 1.49$ g

c) Weight of Pile warp infringe
$= (2 \times 576 \times 30 \times 2)/(100 \times 1000) = 0.69$ g

Weight of pile warp = 183.31 + 1.49 + 0.69=185.49

Weight of weft yarn = $(106 \times 20 \times 58.4 \times 34)/(100 \times 1000) = 42.09$ g

5.24 Characterization of terry towel fabrics

Like other woven fabrics terry towels are also characterized for different physical properties such as abrasion resistance, crease resistance, heat insulation, fabric compressibility, pile withdrawal force, water absorbency, (sinking method, wicking height, water retention, surface water absorption) dullness, bow and skew, dimensional change, colour fastness, laundered appearance and so forth.

5.25 Modern developments in terry production

Terry fabrics are subject to changing fashions: the market is constantly demanding new qualities and designs. The rapid development of electronics with microprocessor controls and highly dynamic stepping motors in combination with modern mechanisms, has enabled fabric designers to produce completely new patterns. One such mechanism is the special terry sley gear with dynamic pile control, as used by Sulzer Ruti in the terry version of the G6200 rapier weaving machine (Fig. 5.23). Through a servo motor, the beat-up position for each pick and thus the type of terry and the pile height can be freely programmed from one pick group to another. In this way, 200 different loose pick distances and hence the same number of pile heights can be programmed in any order desired. For example, three- and four-pick terry, and even fancy types of terry can be combined in the same length of fabric. This gives the fabric designer a broad range of patterning options, and the weaving engineer a technology for improving the fabric structure because the transition from one pattern element to the next can be woven with greater precision. With these elements, Sulzer Ruti specialists have now developed a new patterning method referred to as sculptured terry. At each full beat-up, two pile loops of different heights are formed in the weft direction. The secret

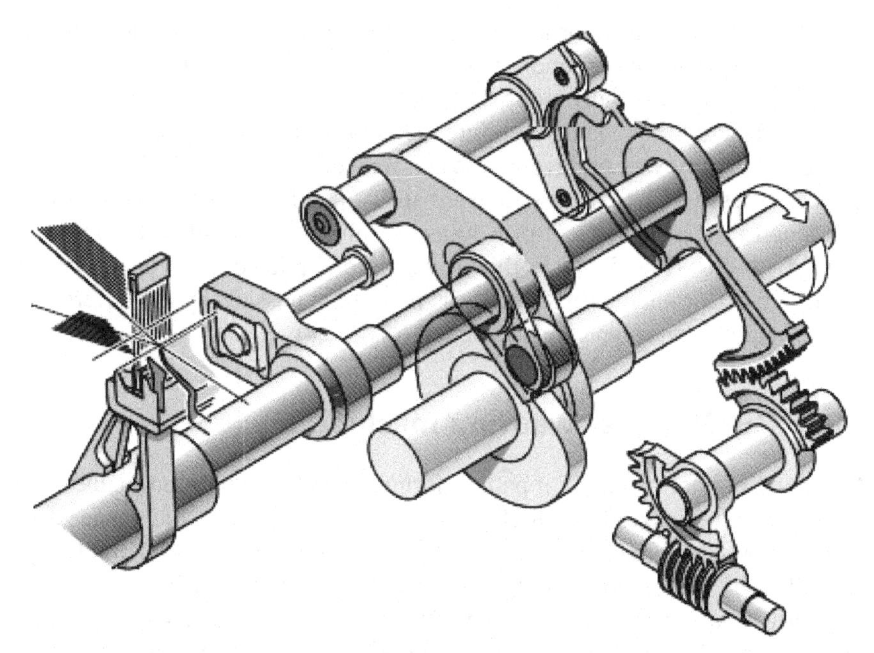

Figure 5.23 The terry sley gear with the dynamic drive

Figure 5.24 Terry fabrics with two pile heights i

of this method of pattern formation lies in the fact that two loose pick groups formed at distances corresponding to the pile heights are beaten up to the cloth fell together. For two short loops, the pile threads are woven into both loose pick groups and for one large loop into the second loose pick group only. The greatest difficulty was to develop a basic weave which results in neat loops without excessive friction between warp and weft at full beat-up. The solution was found in a special seven-pick weave combined with full beat-up at the 6th and 7th pick (Fig. 5.24). In this way, a second pile height is also formed in the weft direction, making sculptured patterning possible by the difference in pile height in warp and weft direction (Fig. 5.24).

5.26 G6200 rapier weaving machine

A precondition for this kind of pattern formation is freely programmable sley travel, as on the Sulzer Ruti G6200 rapier weaving machine. Microprocessor control allows the loose pick distance to be programmed easily and individually for each pick. Adaptations can be carried out at any time, for instance when a pattern is woven for the first time. The terry version of the G6200 rapier weaving machine (Fig. 5.25) can be equipped with a control system for a

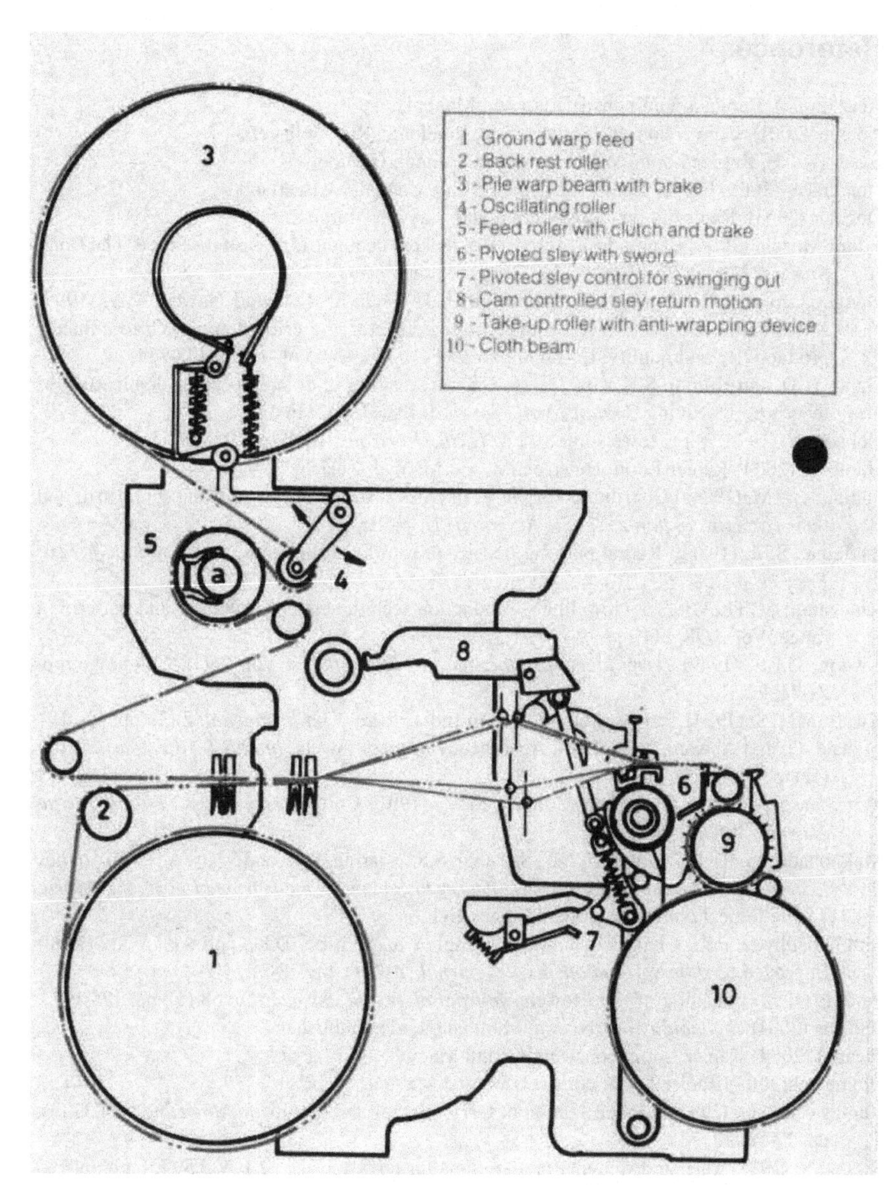

Figure 5.25 G6200 rapier weaving machine

maximum of eight different weft colours or yarns, and a jacquard machine, thus giving fabric designers practically unlimited scope for the design of terry fabrics.

References

Acar (2004), Rapier Loom construction and Manual.

Adanur (2001), Conversion of yarn to fabric, London: CRC publishers.

Baser (2004), Rapier Loom construction and Manual, Germany.

Donaldson (2003) Rapier Loom construction and Manual, Switzerland.

Dornier (2003), Rapier Loom construction and Manual, Mumbai.

Editor Murphi (2006) Bangladesh: Terry towel makers demand 15pc cash incentive. The Daily Star, 1st May 2009, 10:08.

Gangopadhyay, U.K. Vora, H.R. Sakharkar, C.H. Shaikh, R.A. and Gawde, V.A. (1999) Manufacture of terry towel in decentralized sector – a critical approach to enhance productivity and quality. In: *Technological Conference*, Vol.40, pp. 138-144.

Gujar, H D. and Chinta, S.K. (1992), Power loom weaving in Solapur – an in-depth study of terry towel weaving. *Indian Textile Journal*, 103(3), pp. 116-120.

Hobson, V. (1990) Terry towels unraveled. *Textile Horizons*, 10(4), pp. 27-29.

Humpres (2004) Rapier Loom construction and Manual, Zurich.

Ishtiaque, S.M. (1986a) Distribution of Fibres in cross section of rotor and ring spun yarns and their strength. *Indian J. Textile Research*, *11*, p. 215.

Ishtiaque, S.M. (1986b) Radial packing density of rotor and ring-spun yarns. *Indian J. Text. Res.*, *11*, p. 208.

Kienbaum, M. (1978) Terry toweling – production techniques, construction, and patterning range. Vol. 1/78, ITB, pp. 7-27.

Kwatra, G.P.S. (1989) *Terry fabric production – novel methods*. Vol. 36(1). Colourage, pp. 21-22, 27-31.

Kwatra, G.P.S. (1994) Terry towel industry in India. *Asian Textile Journal*, 2 (3), pp. 17-21.

Mahale, G. and Aswani, K.T. (1990) Absorbency of terry towels. *Indian Textile Journal*, 100 (12), pp. 164-166, 169-172.

Mansour, S.A., Saad, M.A. and Mourad, M.M. (1997) Cotton terry fabric. *Indian Textile Journal*, 108 (1), pp. 24-31

Mukhopadhyay, P.K., Chakrobarty, M., Satsangi, S.S., Sharma, D.K. and Vijay, A (1996) *Studies to improve the functional properties of terry towel fabrics made from bi-component yarn*. Technological conference, Vol.38, pp. 121-126.

Mukhopadhyay, P.K., Chakrobarty, M., Satsangi, S.S., Sharma, D.K., and Atal Vijay (1998) Improved terry towel. *Indian Textile Journal*, 109 (1), pp. 18-21.

Patel P. (1998) Finishing of terry towels. *Journal of Textile Association*, 58 (5), pp. 195-197.

Phillips (2001) *Technology of weaving*. Mumbai: Golden publishers.

Picanol (2004) Rapier Loom construction and Manual.

Promatech(2003) Rapier Loom construction and Manual, Mumbai

Ramaswamy, G. (1992) 'Modernization of terry towel weaving. *Textile Magazine*, Vol.33 (3), pp. 17-18, 20, 22.

Rao, I. N. (1998) Yarn and weaving preparations for terry towels. *JTA*, Vol.59 (2), pp. 69-70.

Rouette (2001) Machine manual.

Seyam (2004) Rapier Loom construction and Manual.

Shenai, V. A. (1981) *Technology of Chemical Processing*. Mumbai: Sevak Publications.

Smit Textile(2005), Rapier Loom construction and Manual, Mumbai.

Srivastava, S.K., Chakrobarty, M., Sharma, R.P., Vijay, A and Sharma, D.K. (1997), *Optimization of process parameters to improve functional properties of bi-component terry towel*. Joint Technological Conference, Vol. 38, pp. 121-126.

Srivastava, S.K., Chakrobarty, M., Sharma, R.P., Vijay, A and Sharma, D.K. (1998) Bi-Component Terry Towel. *Indian Textile Journal*, 109 (3), pp. 28-31.

Sulzer R (1998) Machine Manual.

Swani, N.M., Hari, P.K. and Anandjiwala, R. (1984) Performance properties of terry towels made from open end ring spun yarns. *Indian Journal of Textile Research*, 9 (3), pp. 90-94.

Terafdar, N., Chakrobarty, S. and Hossain, M. (2002) Laboratory study for measurement of moisture transport of terry fabrics. Manmade Textiles in India, Vol.45 (3), pp. 98-102

Teli, M.D., Munawar, Q., Chaudhary, S. and Saraf, N. (2000) *Finishing terry towels with softeners*. International Dyer, Vol. 185 (4), pp. 25-29.

Teli, M.D., Munawar, Q., Chaudhary, S. and Saraf, N. (2002) *Finishing terry towels with softeners*. Colourage, Vol.49 (1), pp. 29-34.

Tendulkar, S.R. and Kulkarni, G.N. (1995) Chlorine free single bath bleaching for knitted fabrics, terry towels and yarn. *Textile Dyer and Printer*, 28 (22), pp. 20-24.

Tsudakoma (2005) Sizing for Terry weaving on Rapier loom, Japan.

Venkatpathi, T. M. (1999) Control of fabric defects at grey stage – shuttle looms. In: *NCUTE-Progamme on weaving of shuttle looms*. pp. 85-104.

CHAPTER 6
Weft pile fabrics

6.1 Introduction

These are the compound fabrics characterized by the presence of a smooth fibre surface from one end of the cloth to another end. In pile fabrics, a proportion of the threads, either warp or weft is made to project at right angles from a foundation texture and form a pile on the surface. The projecting threads may be cut or uncut, thus resulting in tufted or looped pile. Weft pile fabrics are composed of one series of warp threads and two series of weft threads, the ground and the pile. Those fabrics are characterized by a short and soft fur or plush pile closely resembling that of silk velvet. This fur-like effect is obtained subsequent to weaving by an operation known as fustian cutting in which certain floating picks of weft are cut or severed by specially constructed knives that are operated manually or mechanically, thereby causing those picks to become more or less erect, thereby exposing their transverse section to the surface which gradually simulates the character of true velvet.

Weft pile fabrics also termed 'velveteens' have the highest pick density of any woven fabric. The picks per inch range from around 175 (60–100 picks/cm) for low- and medium-quality cloth to over 600 (about 200 picks/cm) for the finest fabrics. In order to reach such weft density, the warp setts should be comparatively low (40–90 ends/inch) and the warp yarn has to be kept very taut; due to high warp tension, positive shedding mechanisms are used and the highest qualities of cloth require specially constructed, heavy weaving machinery which cannot operate at high speeds and therefore aggravates further the already done products rates arising out of the high densities of shooting. For this reason, top-quality velveteen's are unpopular. On the other hand, low- and medium-quality cloth in some constructions is very popular and can be produced on standard high-speed automatic weaving machinery using reeds with special deep dent wire.

As already mentioned, velveteens are composed of two series of weft picks, namely pile and ground picks, both thrown from the same shuttle; thus, it eliminates the need of a drop box loom. Pile picks are floated somewhat loosely on the surface, which are to be cut to form pile afterwards, while ground picks interweave more frequently with warp to build up a firm foundation texture to sustain the pile. Thus, if all the tufts were to be withdrawn, there would still remain a perfect foundation texture. This foundation texture is a plain twill or some other sample weave depending on whether a light, median or heavy texture is required.

The pile effect in the velveteens is not produced during weaving but is a result of a cutting operation during cloth finishing. The loosely floating pile filling gets cut but the ground cloth is unaffected by the knife action and thus provides a solid base from which the cut tufts can project and in which they are anchored. The anchorage may be accomplished in two ways: (a) by compression, caused by densely crowding picks of weft and (b) by interweaving pile picks with several warp threads in succession to produce what is termed as fast or lashed pile or by adopting both of these methods. The cutting method differs for different classes of structures and is described together with the appropriate construction. Before cutting the tufts, the cloth is first subjected to a process of liming, in which a thorough coating of lime paste is applied to the face side and then dried. Stiffening the surface float is done in order to define the cutting races more precisely and to ensure crisper cutting. The back of the cloth is coated with flour paste and again dried for the purpose of stiffening it and to prevent the withdrawal of tufts of pile during cutting. After cutting, the cloth is submitted to various finishing processes.

Velveteens are commonly made wholly of cotton but sometimes of rayon. The filling yarn is soft spun, and in better qualities is of combed long-staple

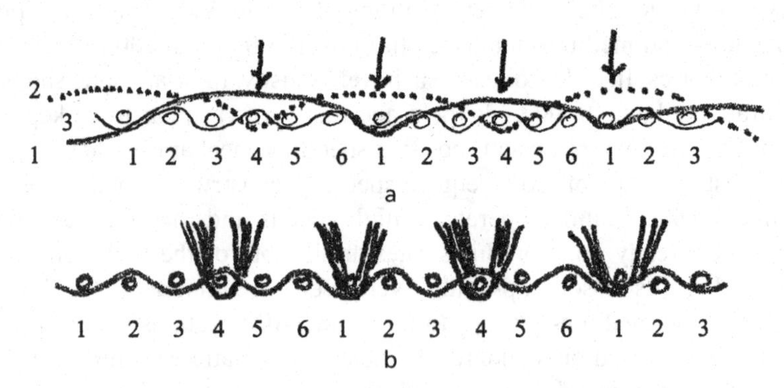

Figure 6.1 a) Principle of formation of long floats; b) Tufts formation following cut

Figure 6.2 (a, b) The concept of the tuft formation

cotton, which when cut gives a soft smooth velvety pile. Velveteen has a wide range of end uses, including sport coats, shirts, blouses, has trimming children wear dresses, lounging pyjamas, slacks and handbags.

Structurally, velveteens may be classified as:

i. All over or plain velveteens in which the surface is uniformly covered by the pile.
ii. Weft plushes – Similar to plain velveteens but arranged to produce much longer tufts and used mainly for upholstery purposes.
iii. Corded velveteens – Also known as corduroys and fustians in which the pile runs in orderly vertical cords of varying width.
iv. Figured velveteens – In which pile figure is produced on the bare ground.

All the above groups may be further subdivided into plain back or twill back structures, depending on the type of weave in which the ground picks interlace with the warp.

6.2 All over or plain velveteens

This class of velveteens has a perfectly uniform surface. The foundation texture being entirely covered by short tufts of pile of uniform length, distributed uniformly over the fabrics. In constructing designs for the fabrics, the chief points to note are:

1. The weaves that are used for the ground and pile, respectively.
2. The ratio of pile picks to grounds picks. These factors together with the ends and picks of the cloth influence the length, density, and fastness of the pile. The interlacing of the pile is almost invariably based either on the plain weave, a simple twill, a sateen, or a sateen derivative.

6.2.1 Plain back velveteens

These are the velveteens produced on plain base and are also called all over velveteens. Following are the various examples in this regard.

Figure 6.3a shows the base pile weave which is plain weave and Figure 6.3b shows the development of the same. This is a simple design repeating on 8 ends and 6 picks. The picks are arranged in the order: 2 pile picks to one ground with pile picks floating over three ends between each intersection. The

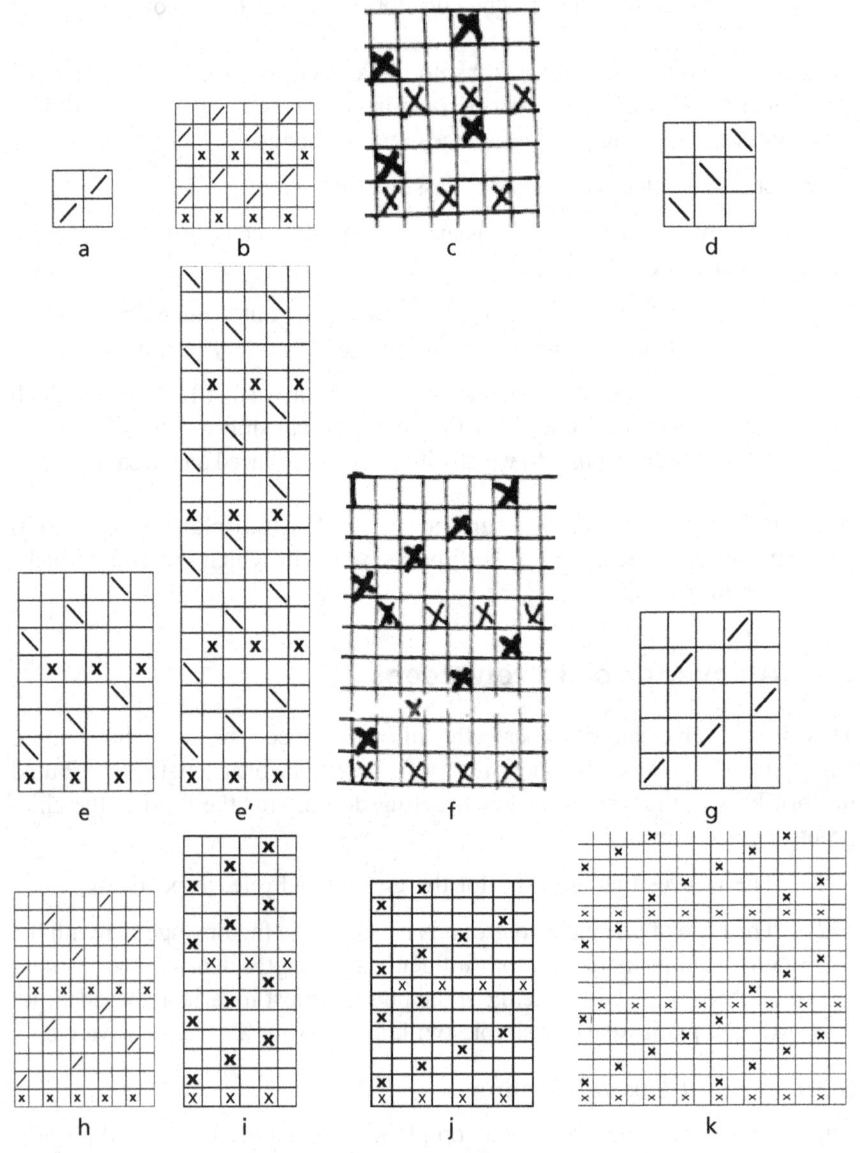

Figure 6.3 (a–k) Weft pile fabrics on plain base

ground picks (represented by 'x') weave in plain order series the weft floats over just 3 ends, the design would produce an exceedingly short and poor pile.

Figure 6.3c: Here also, the arrangement is 2 pile, 1 ground with pile picks bound at intervals of six to cause them to float over 5 warp threads. In this example, every third warp thread is utilized for binding pile picks to the ground texture, namely the first and fourth in each repeat of the pattern, which all warp threads interweave with ground picks to produce the foundation texture.

Figures 6.3d–e′: The base pile weave shown in Figure 6.3d and the same is developed in Figure 6.3e. The reader should note that Figure 6.3e shows the design with ground: pile as 1: 3, whereas the design given in Figure 6.3e′ shows the final developed design for the same base weave but with ground: pile as 1:4 and the repeat size will increase. In fact, this is a method to increase the density of piles will be discussed later.

Figure 6.3f shows the developed pile weave with one-up and three-down twill as base weave. The design is arranged as 4 pile, 1 ground, and pile interlacing is based on the 1 and 3 twill yielding in float of 7.

Figures 6.3g and h: Here, the base weave is 5-end sateen with 5 pile, 1 ground as relative thread of proportion, which gives a float of 9.

Figures 6.3i and j show the modification of Figure 6.3e with 1:6 ground:pile ratio. On the other hand, one can observe that a more modified design is obtained with the same ground:pile ratio when the basic ground is shown on increased number of ends so as to give a longer float . The reader is advised to note the float length in each example as an exercise.

A distinct feature to be noted in the designs is that the pile base weaves are indicated only on alternate ends. The purpose of binding in the pile picks only by the alternate ends is to enable cutting to be more easily accomplished.

Figure 6.1a shows how to interlace picks 1, 2 & 3 with the warp threads. Each pile float stands out furthest from the foundation cloth at its centre. Thus, a series of courses or passage running length wise, termed as races, are formed along each of which (as shown by arrow marks in Fig. 6.1a) a knife passes, thereby serving them in the centre and causing them to become erect on each side of a binding thread to produce the characteristic short tufts of pile as shown in Figure 6.1b and Figure 6.2a.

Although the pile weaves are different, they all will produce the same end result in the finished fabric. Only the length of the pile will be different depending on the length of the float. This is because the pile picks are bound by the same alternate warp threads that are raised for the first ground pick in each repeat of the pattern while the intermediate threads are only raised for the second ground pick, thereby causing the pile picks (2, 3, or 4 or 5 depending on the arrangement) to become equivalent to, and subsequently occupy the space of, only one pick of weft. Hence, all the designs, through having 6, 8, 10, 12, and so on, picks per repeat, are equivalent to only 4 picks in cloth.

In order to produce a dense pile, a very large number of picks are required to be inserted. This can be accomplished because of two reasons:

1. The warp is held under great tension and the ends lie almost straight in the cloth which causes the picks to do most of the bending. This results in the ground texture being formed on weft rib principle; hence, a comparatively large number of ground picks can be inserted.
2. As explained earlier, the system of pile interlacing is such that it enables all the pile picks to be beaten over one another so that each group occupies not more than the space of 1 ground pick.

6.2.1.1 Length of the pile

The length of the pile varies according to the ends of the cloth and the number of ends over which the pile weft floats. To increase the length of pile, reduce ends or increase pile weft float length. To decrease the length of pile, increase ends or reduce pile weft float length with same ends per centimetre.

6.2.1.2 Density of pile

The density of pile varies according to the thickness of weft, length of the pile, and number of tufts in a given space. An increase in thickness of weft tends to make the pile coarser but other things being equal, density increase. A long pile gives better cover and fuller handle than a short pile but fewer number of tufts are formed by each pile pick. Hence, if pile picks is constant, then increased density due to long floats is contracted by the reduction in number of tufts. Therefore, an increase in the length of pile weft float is usually accompanied by the increase in picks.

$$\text{Number of tufts/cm}^2 = \frac{\text{Ends/cm} \times \text{pilepicks/cm}}{\text{Ends in repeat of pile weave}}$$

6.2.1.3 Changing the density of the pile

- Vary the number of picks per centimetre or thickness of weft with the same design.
- Change design to get a different proportion of pile to ground picks. Thus, instead of warp threads binding over 1 pile pick between 2 grounds picks to produce only 1 row of tufts, they may bind over

2 or 3 pile picks to produce a corresponding number of rows of tufts between two grounds picks as exemplified in Figures 6.3i and j.

The pile weave in Figure 6.3i corresponds to that in Figure 6.3e but here there are 6 pile picks to 1 ground pick. In Figure 6.3j, the pile weave corresponds to that in Figure 6.3f but the arrangement is 6 pile, 1 ground. In order to ensure the most perfect distribution of pile, each binding end should hold the same number of tufts between ground picks (as shown in Fig. 6.3i) but this rule is not always observed in practice. The density of pile is sometimes slightly increased by causing additional tufts of pile to occur in certain places only between 2 grounds picks (as shown in Fig. 6.3j) (conversely density of pile may be slightly diminished by omitting tufts of pile in a similar order). In such cases, care should be taken to dispose of additional tufts (or the spaces where taken to dispose of additional tufts in the space where they are omitted) so that they will not tend to develop lines in any direction in the finished fabric. This is achieved by employing a sateen basis instead of the twill base for pile interlacing.

Figure 6.3k is with the arrangement of weft is 5 pile, 1 ground. Here, the successive binding points of the pile weave have been produced in the same direction throughout, then the regular occurrence of the extra tufts tend to develop a series of lines running obliquely across the fabric. In Figure 6.3l, the successive binding points on pile picks are produced in an opposite direction at intervals of two grounds picks, thereby disposing the additional tufts of pile in the 4-end satinette order and in Figure 6.3m, the density of the pile

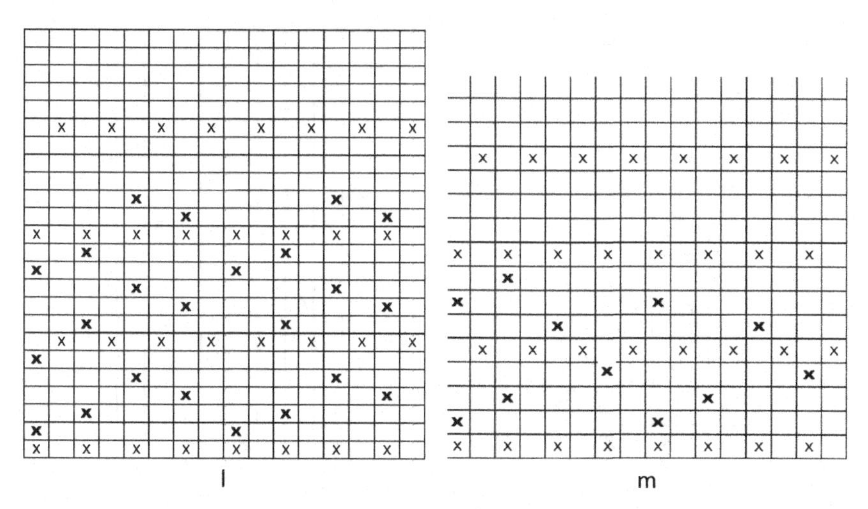

Figure 6.3 (l–m) Plain back velveteens

has been reduced, making use of the same practice of reversing the direction in which binding points are produced at internal of 2 ground picks so as to dispose the gaps. In changing the proportion of pile picks to ground picks, the following points should be considered:

- Increase in density of pile, keeping picks constant
- Changing density, keeping ground structure same
- Same density, different grounds structure

6.3 Fast pile structures

Fig. 6.3n and 6.3o are examples of tabby back velveteens with a fast or lashed pile so called because the tufts of pile are more securely attached to the foundation texture so that there will be no tendency of the tufts fraying out. In all the examples of piles, the tufts are bound in by one end only at a place, and the fastness of the pile is chiefly dependent on the pressure of the picks upon one another. It is, therefore, necessary particularly in the longer piles for very large number of picks to be inserted in order to keep the pile firm. However, the same firmness can be inserted for piles with fewer picks by interweaving the pile picks more frequently and thus making what is termed a fast pile.

Although the binding arrangements of pile picks in Fig. 6.3n and o are different, they will produce no material difference in their end results. In Fig.6.3 n, the warp threads are raised over two out of 4 pile picks between

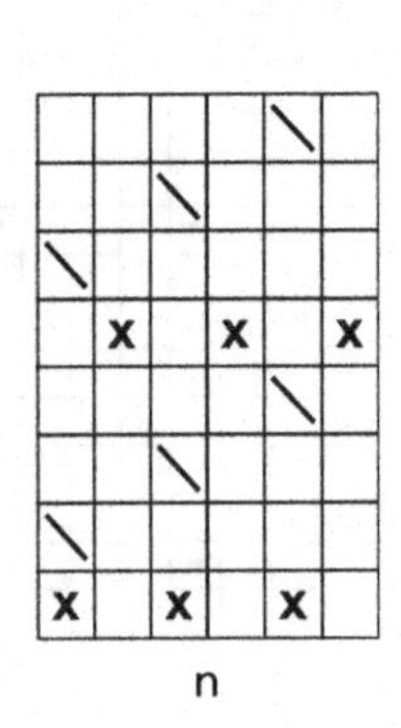

n

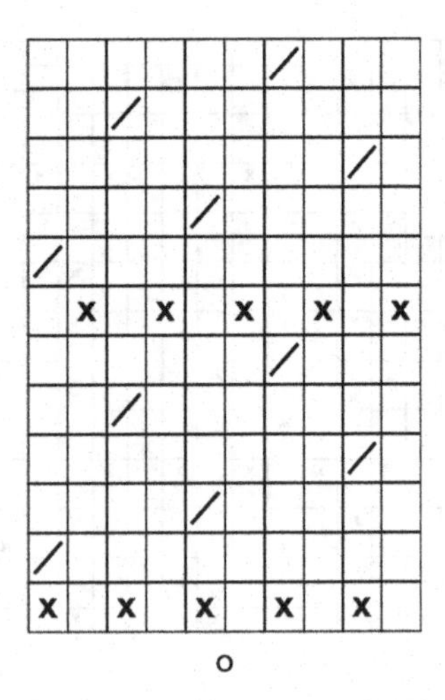

o

Figure 6.3 (n–o) Tabby back velveteens with a fast or lashed pile

the two grounds picks, while in Figs. 6.3 o, the warp threads are raised over 2 out of 4 pile picks between the two grounds picks. In Fig. 6.3o, the warp threads are raised over 2 out of 5 pile picks, such that the pile picks (4 or 5) will occupy the space of two picks in the cloth. This tufts of pile will be formed over an interval of 8 picks (Fig. 6.3n) and 5 tufts of pile will be formed over an interval of 10 warp threads (Fig. 6.3o) from what virtually constitutes 2 picks of weft. It is therefore evident that a fast pile can only be obtained in fabrics of similar quality by sacrificing the density of tufts of pile but there is an advantage of greater firmness giving cloth better wearing qualities.

6.4 Twill back velveteens

Twill foundation is employed in producing velveteens of heavier texture than the tabby back variety. A twill foundation weave is looser than a tabby, and therefore it not only permits but requires a large number of ground picks to be inserted to produce a compact fabrics as well so as to increase the weight and also to hold the pile firmly; otherwise, the free character of the weave would produce a more open texture, thereby permitting easier withdrawal of tufts of pile.

Figures 6.4a–c are velveteens based on 3-end twill foundation which as before is moved on alternate ground ends; the pile picks are arranged in the proportion respectively of 2 and 4 picks to each ground pick.

Figure 6.4 a is the standard design for the moleskin class of fabric, which is usually made in coarse cotton yarns. This is not a pile fabric as the floating picks are not cut but remain in the condition they are after weaving.

If still heavier texture than those that can be obtained with 3 ends twill foundation texture is needed, then a 4-end (2 and 2) twill and other weaves may be employed.

Figures 6.4d and e are velveteens with 2/2 twill foundation texture. Since it enables a very large number of ground picks to be readily inserted, it is employed for the heaviest and dense velveteens. In Figure 6.4d, the pile weave is based on 1/3 twill with 4 pile picks to 1 ground pick, while in Figure 6.4e and f, a 6-end sateen pile base wave is employed with the pile made fast and there are 3 pile picks to each ground pick and in the Figure 6.4f show the corded velveteen with twill as ground.

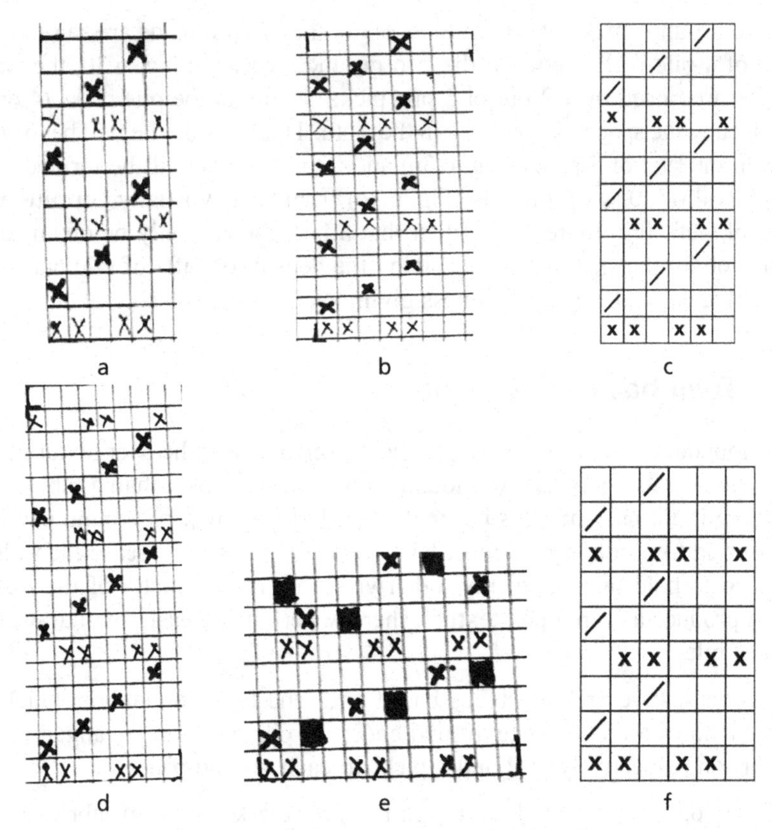

Figure 6.4 (a–f) Twill back velveteen

6.4.1 Designs to simplify the cutting operation

In order to reduce the time occupied in cutting of pile, the pile base waves are indicated only on every third end; therefore, only one-third as many longitudinal traverses of the cutting knife are required as there are ends in the width of the cloth. The distribution of the pile, however, is not so perfect and the surface of the cloth has a coarse appearance.

6.4.2 Qualities of all over velveteens

Quality of any velveteens construction can be varied considerably by changing the warp weft yarn setting and counts. The reduction in picks is usually compensated for by the increase in ends if similar density pile is required. A feature of these fabrics is that there is considerable shrinkage from reed to cloth, which varies from 12.5% lighter velveteens to 20% heavier ones.

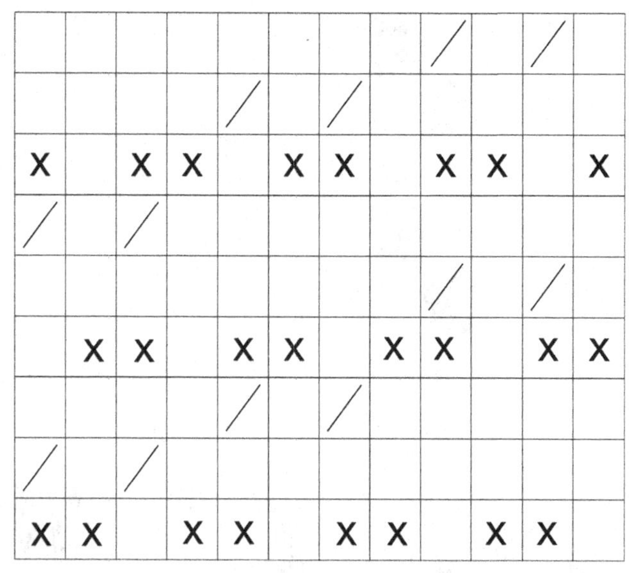

Figure 6.5 Weft plush

6.5 Weft plushes

Weft plushes (Fig. 6.5) are simple velveteens in which the pile weft is allowed to float over a considerable number of warp threads to produce a longer pile, the tufts of which are more firmly interlaced by interweaving them under and over three of five consecutive warp threads. In all other respects, their construction is similar to that of ordinary velveteens fabrics having short pile. Weft plushes are of heavier weights and are chiefly employed as upholstery cloths.

6.6 Ribbed or corded velveteen

Corded velveteens (Fig. 6.6) are velveteens with ribs or cords produced length wire or parallel to the warp. Ribbed velveteens are of two kinds (a) ribbed velvet or velveteen and (b) corduroy or velvet cords.

6.6.1 Ribbed velveteen

These are also known as hollow-cut velveteen. These are woven as ordinary plain velveteen and afterwards made to assume a corded appearance by a special method of cutting in which a cutter first passes a knife along certain races in each cord with the blade vertical as in ordinary cutting and then along

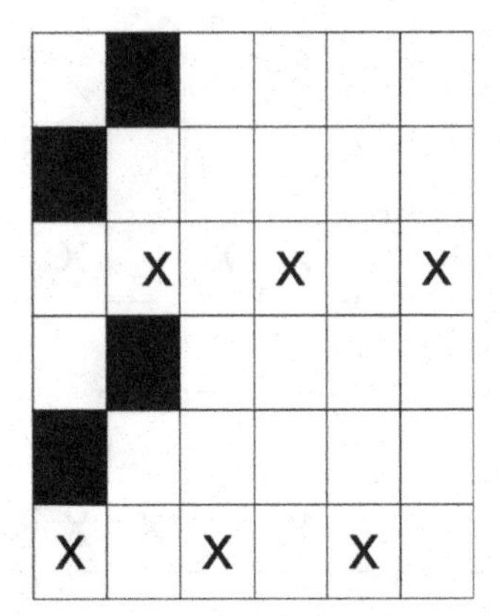

Figure 6.6 Corded velveteens

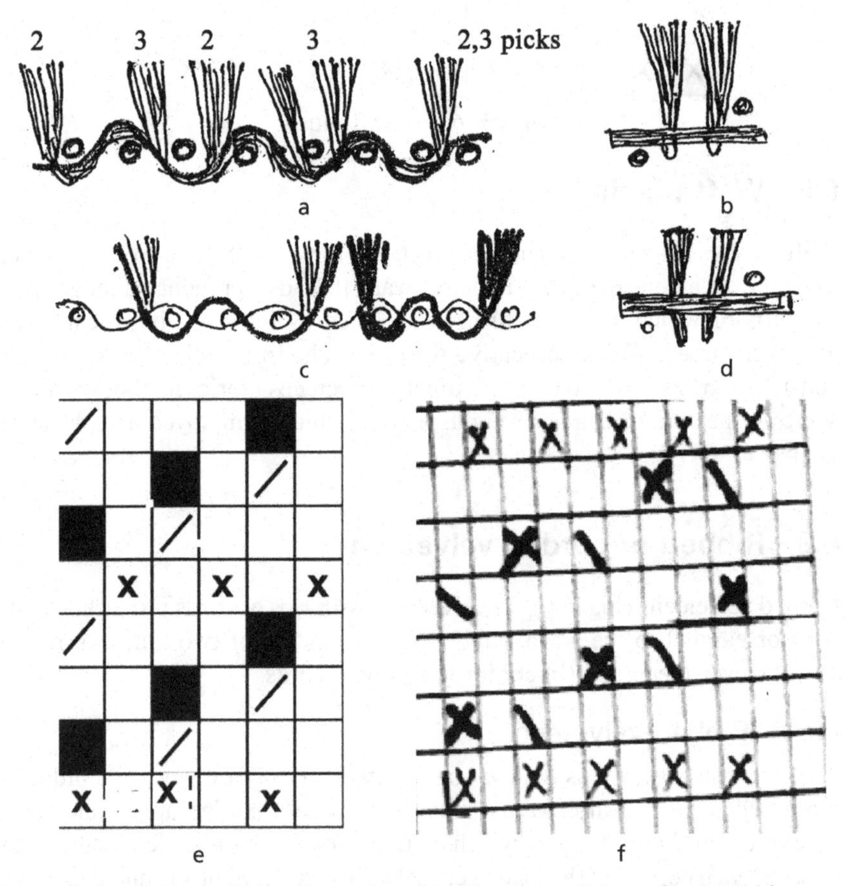

Figure 6.7 a) Fast pile cross section; b) side view of the tuft; c) Making the pile fast; d) End view of the fast pile; e) design for the past pile on twill; f) Fast pile with ground on five ends

intermediate races with the knife blade held at different angles to 7 floats of weft out of the centre and so form longer and shorter tufts which develop rounded ribs of pile. Figure 6.8 shows how the different lengths of tufts join together to give a rib effect.

The width of the cords is not regulated by the number of warp threads on which the woven design repeats but is arbitrarily decided by the cutter, who may produce various widths of cords from exactly similar fabrics.

6.6.2 Corduroys

Callaway Textile Dictionary defines corduroy as 'a stout durable cotton fabric having pronounced cords running warp wise'. The cords are made up of cut pile and are termed wales. They are separated from one another by narrow cut lines known as welts. The wales per inch may vary from 1 to 22. The material is woven with two systems of filling, ground and pile and has a high number of picks per inch. In the finishing of corduroys, the pile picks are cut to form the tufted cords. A very satisfactory fabric for outdoor wear of all kinds are sportswear, slacks, children's clothing and in the fine wale types as a dress fabric. In French, *corde du roi* means 'kings cord'.

In these structures, the pile picks are bound in at intervals (according to the width and character of cord required) in a straight line. The cuts are made right at the centre of the space between the pile binding points with the result that the tufts of fabric project from the foundation in the form of cords or ribs running length wise of the fabric (Figs. 6.6, 6.8, and 6.9).

Corduroys are usually of heavier textures than ordinary velveteens and the foundation texture is usually a twill weave or any other simple weave. The pile and ground picks are usually arranged in the ration of 2 pile:1 ground.

Sometimes, lighter textures of corduroy are produced when plain weave is employed for the foundation texture with 3, 4, or 5 pile picks for every ground pick to get a denser pile. These finer cords are suitable for boys and ladies dress material.

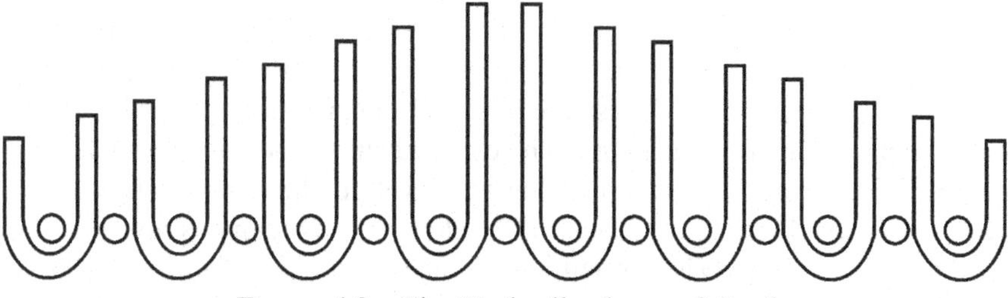

Figure 6.8 The Umbrella shape of Cord

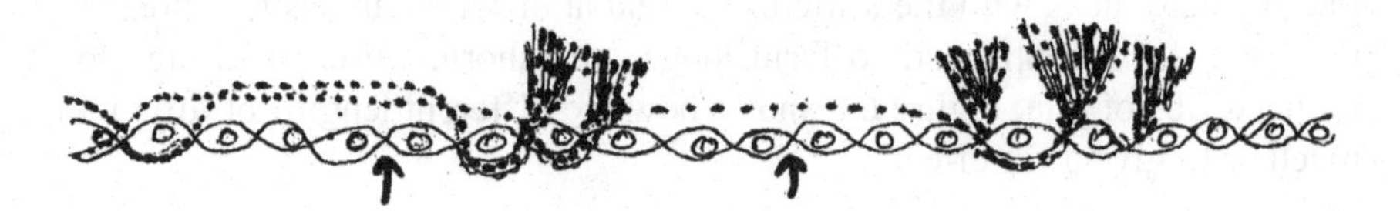

Figure 6.9 Cross section of corduroy

1. Corduroy with 2 pile:1 ground pile picks bound in plain order on two consecutive ends. The plain binding weave of the pile picks reversed in alternate cordswhere design extends over the width of two cords, and each pile pick forms alternate long and short float.
2. Corduroy with 3 pile:1 ground pile picks bound in plain order on the consecutive ends. Here, binding is the same in each cord. In this case, all pile floats are equal.

The result is practically same whichever method of binding is adopted because the floats are cut in the middle of the space between the binding points, consequently, in either case, one side of each tuft is longer than the other side. The difference in the lengths causes the ribs to have a rounded formation as the long side of the tufts form the centre and the short side the outer parts of the cords. Simplest and smallest design of corduroy fabric, commonly termed thickset cord repeats on 6 ends × 9 picks and has a foundation texture based on the 3-end twill weave, with 2 pile picks to 1 ground pick. The floats of weft are very short being over only 3 warp threads – thereby producing a short stubby pile, the tufts of which are firmly bound in the ground cloth, after the manner of lashed pile described under velveteens.

6.6.2.1 Corduroy cutting

The pile wefts of a corduroy fabric can be cut by hand with a fustian knife, but a more popular method is to cut the piles with mechanical knives. The machines are of two distinct types and are called circular knives and straight knives of which the former is more popular.

Figure 6.10 illustrates a circular knife machine. A number of sharp-edged steed discs 'B' are mounted on a mandrel A placed across the machine and revolving at a high speed. These are adjusted to suit the width of the cords, one disc placed against each cord. The cloth advances the knives and as they are revolving, there is cross rail C over the bevel of which both falls down at right angles. At this point, the races of the cloth are placed in position with the knives. The wide wire D is inserted in each race as shown in Figure 6.10. These wires guide the floats of weft to the wires, keeping the knives in the centre

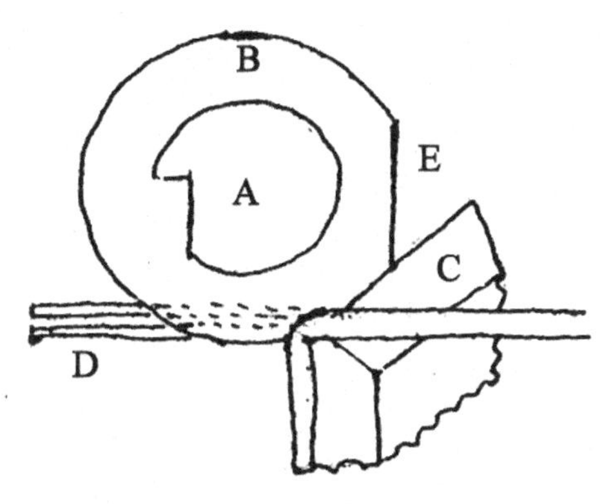

Figure 6.10 Corduroy cutting

of each rare and also keeping the floats stretched while cutting. A segment is cut off of the disc, reducing the diameter of the knife at that portion. This portion is brought parallel for both, insertion of it or taking it out after cutting.

The wires are pushed intermittently by means of a suitable contrivance as the cloth advances to wide a fresh portion of the cloth for cutting.

6.7 Figured velveteen

Velveteens are ornamented by embossing designs or by developing design with the help of a jacquard machine. In velveteens figured by jacquard, the design is developed with piles and the ground portion is a specially constructed lower plane.

When the pile picks are not forming piles in certain places, they are taken to the back of the grand portion and are made to bind with the warp threads in a way which is opposite to that of the pile portion. The surplus pile weft is sometimes allowed to float at the back of the ground portion or is made to interweave with a fine extra warp to produce a light gauge fabric. In both these cases, the surplus material is cut away.

6.8 Figured corduroy

Similar to figured velveteens, corduroys are also ornamented with figures developed by jacquards, and so on, pile wefts are floating on the face of the fabric where figures are required when it is traversing through the ground portion, and it is interweaving with warp in a way opposite to that of the figured position.

CHAPTER 7

Gauze and leno

7.1 Introduction

The terms gauze or leno are used to describe a separate class or group of the weaves and not types of fabric. The characteristic feature of these weaves is the partial crossing of two or more ends in a fabric, caused by being pulled out of their normal straight and parallel course, first to one side and then to the other side of other warp threads which cross and across in some definite order, that is why these weaves are sometimes referred to the term cross weaving. Cross weaving is a useful principle of fabric structure which is adopted extensively in the production of silk cotton, worsted, and linen textured for almost every variety of purposes, such as garments that are employed for as under clothing, blouses, curtains, shirting dress material, and a host of other domestic articles. The weaves are also employed for producing fabrics meant for industrial use such as filter cloths, screens, and sieves.

The crossing of ends from side to side performs two useful function

1. It gives added strength to the fabric – if two fabrics are woven with the same construction (ends and picks/inch) same yarn number, same twist, and so forth and one has a leno weave while the other has any other weave, the fabric with the leno weave will be the stronger.
2. It gives stability to the fabric – after repeated laundering, the leno woven fabric will look reasonably fresh and crisp, while the same weight and construction fabric if woven with plain weave, will appear ragged. The ends and picks will be distorted and thick and thin places will show.

7.2 Types of threads used in cross weaving and RTP

Two series of warp threads and one series of weft threads found used in cross weaving. They are standard end and crossing end in the warp direction and regular weft or ply of weft (Madras gauze). Thus, the two series of warp threads are, respectively, known as standard or standing or regular ends and crossing ends or doup ends. The standard ends lie almost in a straight line and crossing ends are pulled out from their normal, straight, and parallel order first to one side of the standard ends and then to the other side. RTP of standard: crossing is generally 1:1, but sometimes it may be 1:2, 1:3, 2:1, 2:2, 2:3, and so forth depending on the requirement in design which is related to end use or application.

7.3 The principle of cross weaving

The terms gauze and leno denote distinct woven effect, but they are indiscriminately used, and is therefore essential to know how they differ from each other but in spite of their difference, they are still based on the principle of cross weaving. Hence, before knowing the difference between them, let us see what is common in them. The simplest example of cross weaving is in which one or two warp threads cross each other on successive picks, or pairs of picks in regular succession so as to produce an open net-like structure of uniform texture.

7.3 Beaming drafting and denting

It all depends on the type of structure such as simple gauze or modified and according to the beaming varies. For example, in the case of simple gauze, there will be only one beam such that both the threads are subjected to same tension, and if there are one standard and two crossing, then two beams are required one for standard and the other for crossing. As far as the draft is concerned crossing end is shown with easer bar, crossing heald, and doup wires, whereas standard end has standard heald. The denting plan as found in other class of fabrics is equal to RTP in the warp.

7.4 Position of standard with respect to crossing

The position of crossing with respect to standard is described by top and bottom douping. If the crossing end is under the standard end (which is the normal practice), it is referred as 'bottom douping' and if the crossing is arranged over the standard it is known as 'top douping'.

7.4.1 Bottom douping

Figures 7.1a and b represent cross-sectional views of two successive sheds formed. Figure 7.1a (situations 1 and 2) show the crossing ends (black) forms

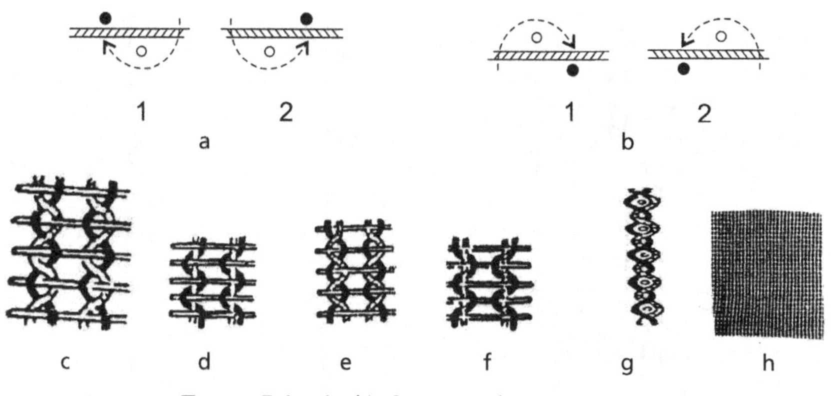

Figure 7.1 (a–h) Gauze and Leno weaves

the top shed on the left of the standard ends (white) and on the right of the standard ends. Thus in simple gauze, the crossing and is up and the standard end is down on every pick but in between each successive shed the crossing and crosses under the standard end prior to each lift and the weft is held between the half twists of the crossing end (Fig. 7.1c to g).

As bottom douping is mechanically more convenient and is much more commonly encountered and most of the examples given subsequently have been worked out on bottom douping principle.

7.4.2 Top douping

On the other hand, if the crossing end is above the standard end, it is referred as 'top douping' at b (situations 1 and 2). It is noted from the cross sectional views that two successive sheds are formed in which crossing end forms the bottom shed to the right and to the left of the standard end, the later on each pick remains in the top shed. Thus, in top douping, the crossing ends are down and the standard ends up on every pick but as the crossing end is transferred alternatively from one side of the standard to the other between each pair of the weft is held securely in the half twists of the crossing end (Fig. 7.1c to g).

7.5 Position of crossing and standard ends

It is to be noted that two ends can be arranged either in left or right side of the other, that is, Figure 7.1c show crossing is left and the standard is towards right and Figure 7.1e show standard at left and crossing to right of standard. Further, if both ends are drawn from the same beam (production of gauze) they appear as shown in Figures 7.1c and e. But if standard is tensioned or

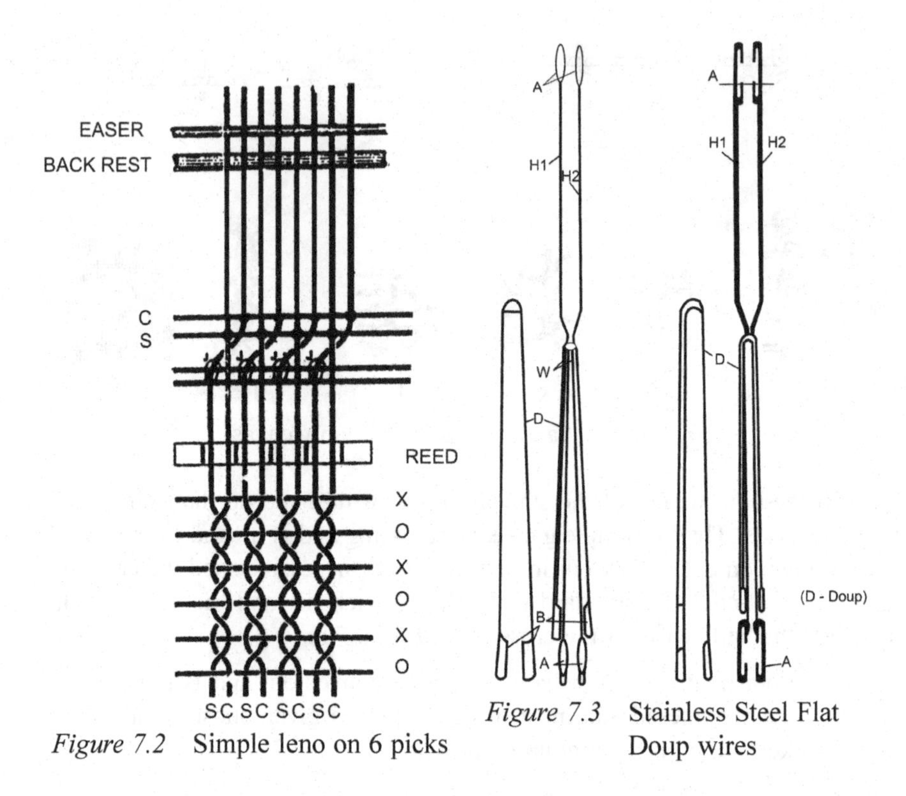

Figure 7.2 Simple leno on 6 picks

Figure 7.3 Stainless Steel Flat Doup wires

straight, the crossing will tend to cross and again the positions may be either to right or left as shown in Figure 7.1d and f. However as mentioned above, in all the cases the weft or pairs of weft will be held in between the ends or pairs of ends (as sometimes there may be two crossing + one standard or two standard + one crossing or two crossing + two standard and so forth).

A sample of a gauze fabric is shown in Figure 7.1h, indicating high porosity of the fabric and its suitability in medical applications as bandage cloth or in furnishing trade as curtain cloth. The arrangement of crossing or figuring and standard or regular in the form of draft, denting plan and hence the design is shown in Figures 7.2 and 7.3.

7.6 Loom equipment necessary for cross weaving

Leno weaves may be woven on cam looms/dobby looms/or looms equipped with a jacquard head. To weave leno on a cam loom or dobby loom, several extra attachments may be necessary such as a jumper motion, an easier or slackener bar and a yoke, the functions of which are explained at the end of the chapter.

7.6.1 Harness heddles

Conventional harness frames may be used. On the first two harness, special healds called standards are employed They are made as male and female, to make a pair. Two pairs of these standards are needed to accommodate a doup needle. This group of five pieces (two male standard, two female standards, and a doup needle) are needed for each complete standard doup. On the other harness, conventional heddles are used.

7.6.2 Facilities for cross weaving

1. Cotton knitted healds
2. Flat steel doup wires (with and without slots; slots at one end and slots at both end)
3. Flat steel doups with an eye
4. Eyed needle and slider frame devices
5. Rotating bobbin and geared disc mounting

Cotton knitted healds

Figure 7.4c shows the formation of three sheds using cotton knitted healds. Handloom sector prefers this set-up.

Flat doup wires

These are used in the power loom sector and different varieties of stainless steel falt doup wires namely flat steel doup wires without the slot, doup wires

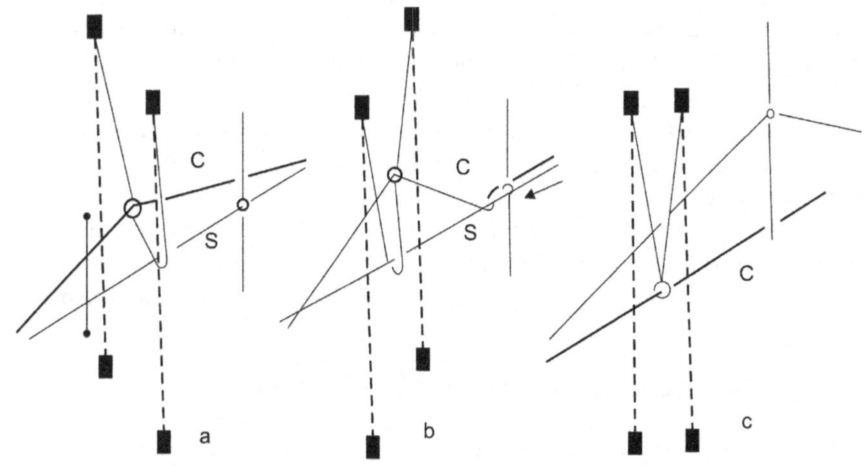

Figure 7.4 (a–c) Sheds formed in Leno

with slot at one side only and flat steel doup wires with slot at both the ends are found. Accordingly, the arrangements in draft and peg plan will vary. The reader is directed to take a note of these aspects as he approach further. Figure 7.4b shows flat steel doup with and without slots. The doup wires consist of two legs namely H_1 and H_2 carrying provision for mounting at A and a doup D winked at W and with B provision for mounting.

What is the purpose of slots in doup wires?

By providing slots, it is possible to manage the position of crossing end as and when required to form the lower line of shed without operating the legs H_1 and H_2. Owing to the presence of slot, the crossing end will slide through the slot when the crossing heald is down. Hence, there is no need to bring H_1 and H_2 down.

Number of harness

It takes minimum three harnesses to weave a plain leno. However, four harnesses are commonly employed: 1st standard (harness), 2nd standard (harness), jumper or ground harness and back harness (of these if necessary, back harness is the one that can be eliminated) As mentioned above, the need for several extra attachment to the loom, the functions of these are

How to represent healds in draft and lifting plan in leno weaving on point paper?

On point paper, one has to note the following method for representing leno. For drafts (with any one system of douping and normally bottom douping is preferred and accordingly show standard end above the crossing end) use one heald per crossing end (Marked by C & connected to easer- E) and one heald for regular or standard heald (marked by S) and doup wires H_1 and H_2. Based on RTP, the C:S is shown in the same or single split in reed and design includes the system of douping and accordingly the position of crossing is indicated with respect to standard between each picks. However, it is to be noted that while drawing design of leno, care must be exercised to show the weft in between two threads. The peg or lifting plan is shown on each pick and it refers to the type of shed formed (the respective elements such as H_1 or H_2 or E and so forth, of the shed based on the type of drop wire used) and is indicated by writing 'x' in the intersection point of respective heald and pick.

7.7 Basic sheds formed in gauze and leno

Gauze and leno production involve the formation of three basic (Fig. 7.5 to 7.7) sheds namely open shed, crossed shed, and plain shed.

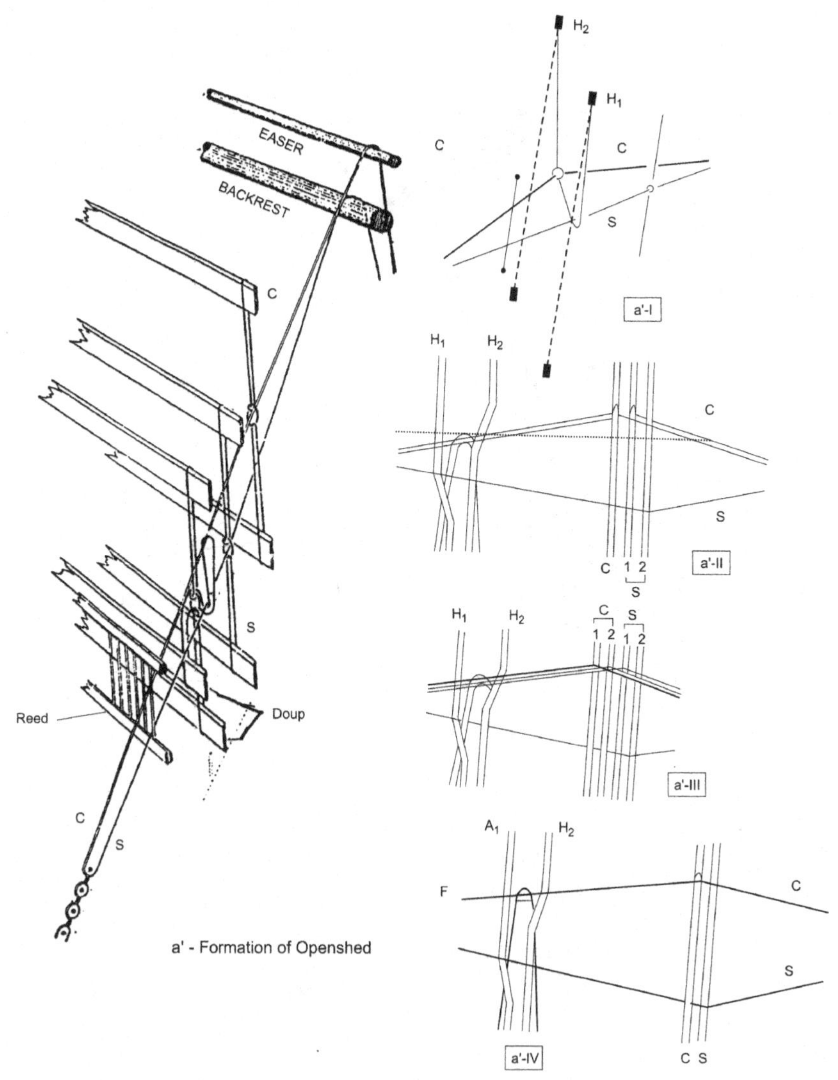

Figure 7.5 Open shed formation

7.7.1 Open shed

A straight or open shed is one in which warp threads are separated without deviating from their normal parallel course. The formation of the open shed with a bottom doup harness is affected by raising both the crossing (which controls doup warp) and the doup over standard. In other words, when crossing end forms the top layer of the shed, it is known as open shed. During this

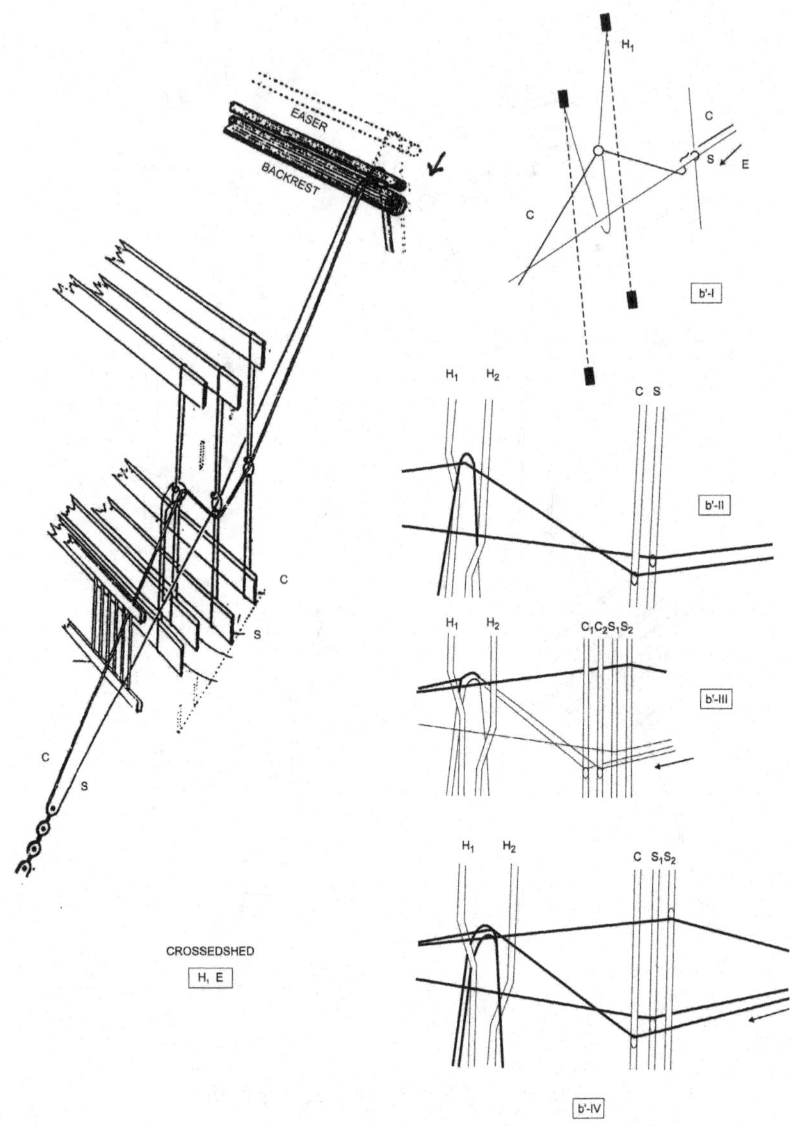

Figure 7.6 Crossed shed foramtion

shed formation, no easer bar or slacker device is operated. Moreover, if more than one crossing ends are used, then at least one will form top shed and the remaining will settle as the bottom layer. The draft consists of separate heald shaft for crossing end and standard end. Figure 7.5a shows the formation of the open shed and the same is the arrangement with conventional doup set

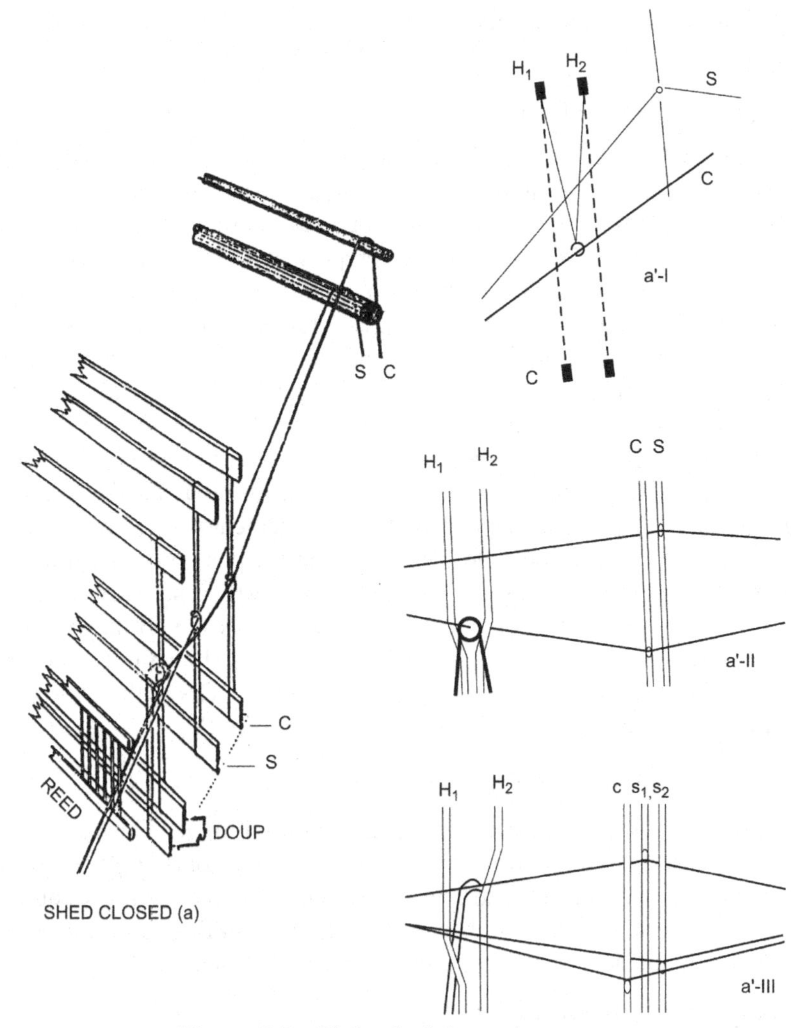

Figure 7.7 Plain shed formation

as shown in Figure 7.5a′. Formation of the open shed with different facilities of crossing is shown in Figure 7.5a -I to a′-IV. A line sketch of the open shed show (Fig. 7.5a′-I) the crossing end forms the top layer of the shed and standard at the bottom of the shed and this is formed as long as the crossing end appears on its 'right side' (not to be confused with direction) instead of 'wrong side'. Hence, operate H_2 C in all such cases. Figure 7.5a′-II show the situation with two standard ends and one crossing end using flat steel doup wires with slot one side only (here the need or presence of slot is explained as it is not necessary in open shed with one crossing end). Figure 7.5a′-III show 2:2 for C: S and 1:1 is shown in Figure 7.5a′-IV.

7.7.2 Crossed shed

A crossed shed is one in which doup warp threads are raised on the opposite side of standard warp threads or the crossing end even though forming a lower line of the shed by the drop of crossing heald is lifted to top by crossing it over the standard. The crossing is accomplished by the operation of easer and hence in all the crossed shed, the arrow mark show this aspect. It should be noted that some extra length is necessary for the crossing end as it is at the bottom to cross comfortably. But to reduce the tension on the standard end when the crossing is lifted by crossing the standard, an additional device is operated known as 'JUMPER'(about this element reader can find information late part of this page) when repeatedly it is necessary to form open and cross sheds alternately to produce gauze fabric. The formulae for the crossed shed is operated. H_1 E Figure 7.6b′ show the arrangement with conventional doup set. The arrow mark show clearly the shift in the position of easer and hence slackness is created along with fast let off of warp yarn. Figure 7.6b′-I show the set up with flat steel doups. Figure 7.6b′-II show the set up with flat steel doups without slots and 1:1 RTP, a 2:2 RTP, C:S is shown in Figure 7.6b′-III (flat steel doup with slot at one side only) and 1:2, C:S with flat steel doups with slots at one side only is shown in Figure 7.6b′-IV.

7.7.3 Closed shed and plain shed

Figure 7.7a show the closed shed which is needed to mend any end breakage during weaving. When the standard end form the top line of the shed and crossing is allowed to settle down, it's known as plain shed. Sequentially, Figures 7.7a′-I, a′-II, and a′-III show the formation of the plain shed with doup wires without slots, with slots when 1:1 and 2:2 C:S is used.

How gauze differs from leno?

The answer is quite bothering as both are similar in all the respects and terms are indiscriminately used; however, one can say that if a plain shed is formed then the construction or structure is known as 'leno' otherwise it is identified as 'gauze'.

Simple gauze production

Figure 7.2 shows the arrangement of threads in simple gauze. Four pairs of threads are shown wherein each pair consists of standard and crossing threads. One of the features of simple gauze is that both the threads (1:1) are drawn from the same beam and are crossing equally and sequentially to form open perforated structure. Figure 7.2 also show the draft, denting, and design of simple gauze. The application of this design is found in mosquito net fabric production and bandage production. Of late, presently using PP, transportation bags for vegetables are found rapidly accepting by the market.

How is crossing accomplished?

During shed formation, the respective healds will be lifted as per the requirement and in all the cases except for cross shed, easer will be in its normal position and let off will deliver the regular delivery rate of warp yarn. But on crossed shed, first the easer bar shifts its position and at the same time the let-off beam unwinds or lets off the extra length of yarn so as to facilitate the crossing. Actually, the crossing end will be at bottom of the shed during crossed shed and will be lifted to form the top line of the shed by doup ($H_{1)}$ and due to the motion of H_1 and easer, crossing end cross the layers. Of course, in order to reduce the strain on standard end on the continuous open and cross sheds formation, Jumpers are put in action.

7.8 Different types of structures in gauze and leno weaving

By varying the RTP, C:S and number of easer bars different structures can be produced (with bottom douping) as listed below

1. Simple leno on 10 picks using ordinary flat steel doup wires
2. Leno on 10 picks with single side slotted doup wires
3. Leno on 10 picks with various effects using 2:2 C:S RTP (using slotted doup wires) Figure 7.10a to e
4. Leno on 10 picks with 2:2 C:S, two easer bars and double-slotted doup wires Figure 7.11a to e
5. Net leno
6. Russian cords

7.8.1 Simple leno with ordinary doup wires

Figure 7.8a to e represents the design, draft, peg plan, and respective sheds formed in the production of simple leno on 10 picks using ordinary doup wires. Here, 1:1 C:S and a single easer is used. The first pick is a crossed shed as the crossing end is appearing on its ‘wrong side’. The second pick forms open shed as the crossing is on top of shed line or appearing on the same side or ‘right side’, followed by the plain shed on the third pick where in crossing end

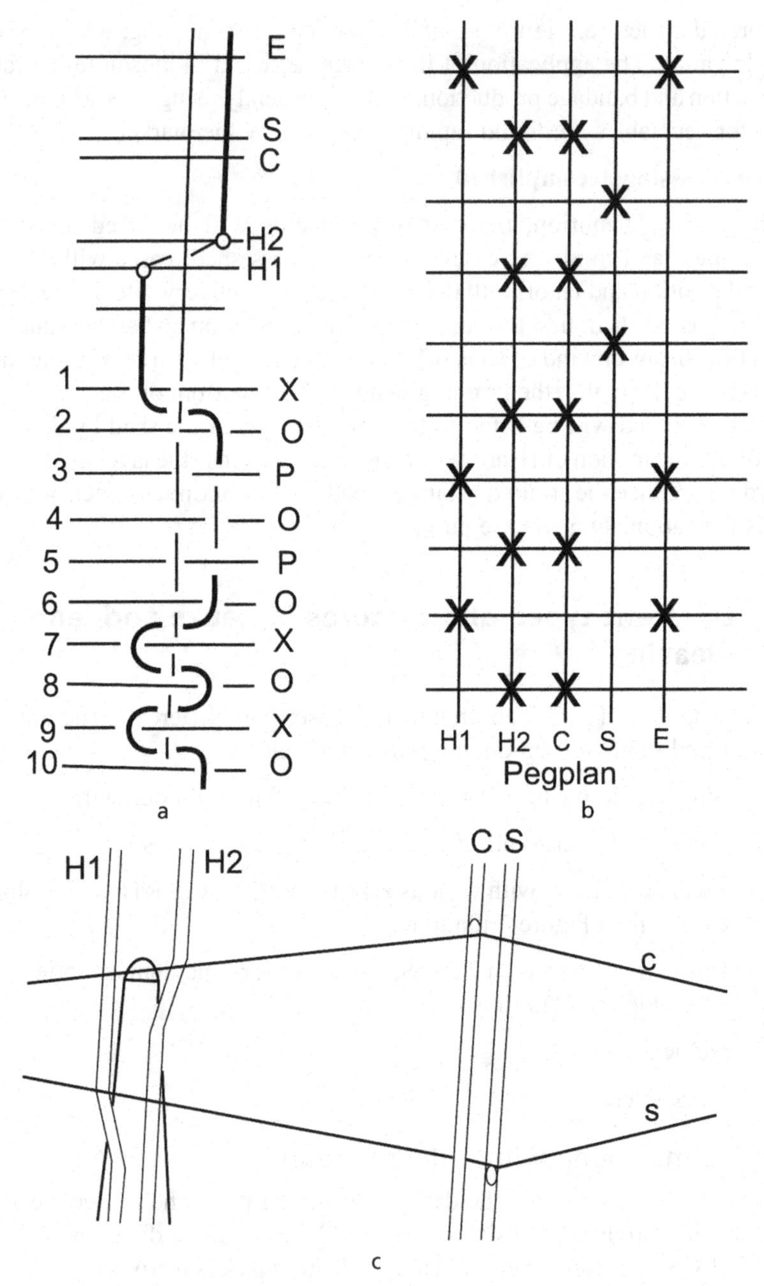

Figure 7.8 (a–c) Simple leno with regular doup wires

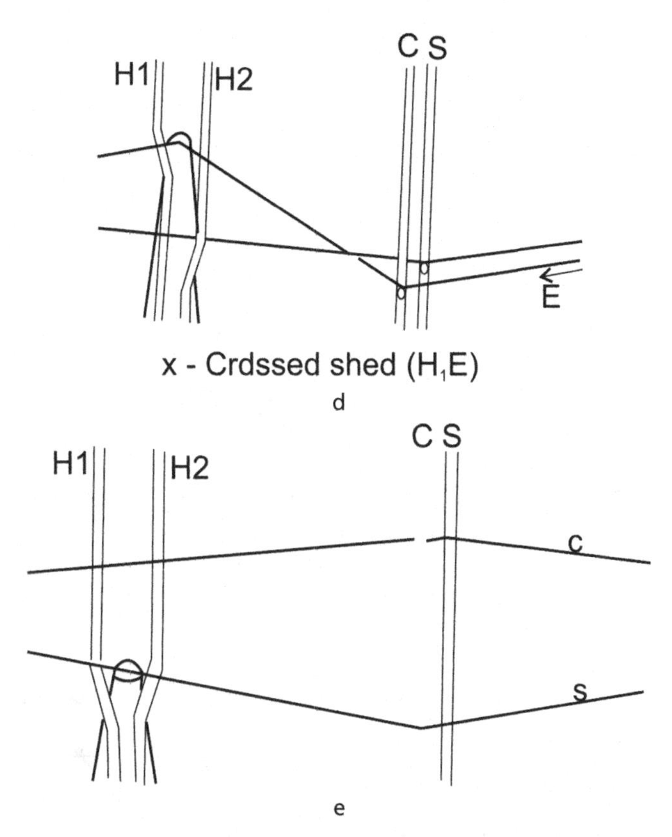

Figure 7.8 (d–e) Crossed shed lines

is down and the standard end is top. Now referring to golden rules like for cross shed using ordinary doups, operate easer with H_1 and for open shed lift H_2 C. For plain shed, lift standard heald and so forth accordingly, the crosses are marked on each pick in peg or lifting plan.

7.8.2 Leno on 10 picks with single side slotted doup wire

Three sheds formed with 1:2 C:S is shown in Figure 7.9a to e in which leno is produced with steel doup wires having slots at one side only. As mentioned earlier, the crossing end forms the lower line of the shed through sliding. Peg plan is constructed on the guidelines as explained in the previous example.

Note: Reader is directed to analyse the remaining examples.

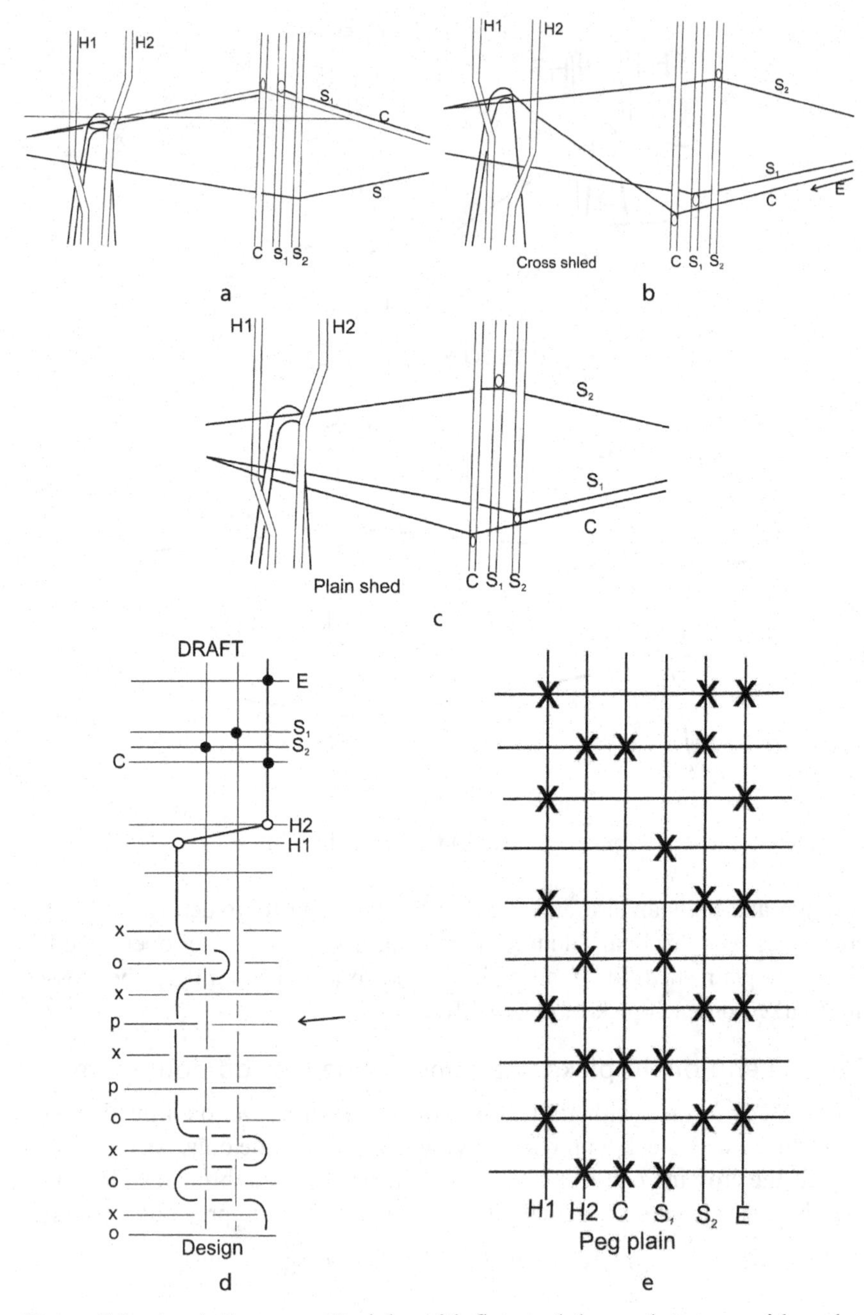

Figure 7.9 (a–e): Leno on 10 picks with flat steel doup, slot at one side only.

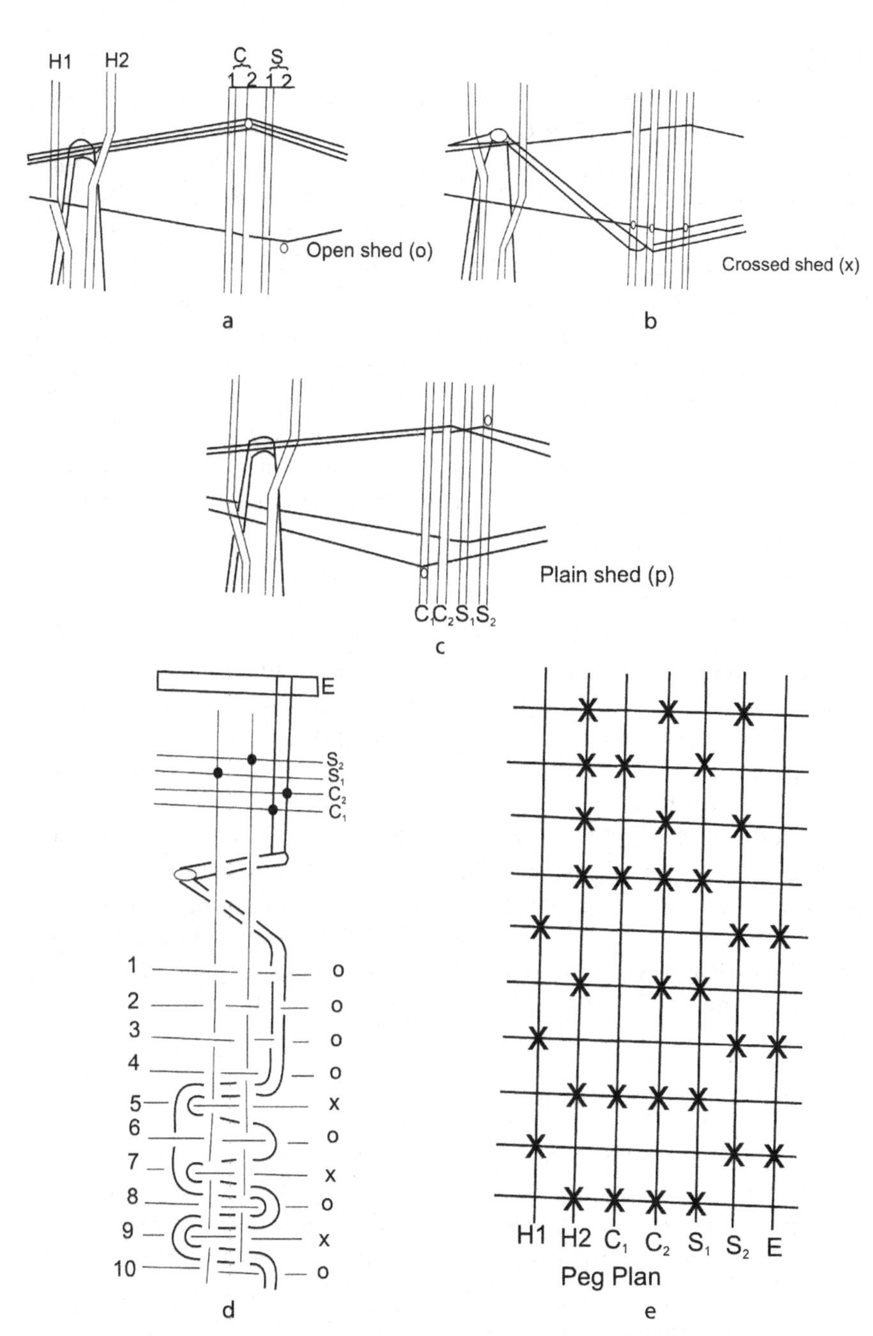

Figure 7.10 (a–e) Leno with 2: 2 arrangement

7.8.3 Net leno

This is also known as point or counter leno wherein point draft is used. Here, two easer bars are used with S: C as 2:2. One set of crossing ends are controlled by easer bar and the other pair working will be arranged in a staggered manner more or less like half drop basis. Standard ends weave plain and are drawn in V form. In a pair of crossing end, one end will cross to the left side and the other to the right side and both will unite to float over few picks prior to shifting from respective sides. Two sets of crossing ends are drawn from separate beams to permit full sideways movement or deflection when the standard ends are straight. The terms net or spider leno is commonly applied to those styles in which the crossing ends are floated on the surface and are interlaced to form wavy patterns. The effect is obtained when crossing ends of some special material as compared to the ground material. On the other hand, ground yarns are of compact construction across which the crossing end cross. Openness is due to missing dent effect.

7.8.4 Russian cords

Basically, this is a plain-based structure produced by 4:1 RTP S:C such that fine standard material will allow the coarse crossing end to cross. Plain weave is arranged in between the standard and crossing ends and form open structure. The weft in the structure is concealed, the surface formed by a distinctly bulging cord effect. The fabric is woven with face down concept on bottom douping. Use of shaker or jumper is recommended as alternatively open and cross sheds are formed. Further ornamentation is improved by using heavy wefting densely packed giving a wrapper effect. Heavy cord lines are thus produced which on account of the contrast in colour with the ground warp and weft appear to be formed in extra weft. The crossing end should be very long and its length varies in relation to reed and pick of fabric. In another approach, standard ends are coloured as compared to the crossing.

7.9 Easer or slackener

In the open shed, the tension on both the grounds and doup ends is approximately the same. On the crossed shed, the doup end requires more yarn than the ground end in order to make the shed. Unless there was some way of easing or slacking this doup end, it would break under the extreme tension. This easing or slacking is accomplished in one of two ways: (a) positive slackener (b) negative slackener The negative slackener is the simplest device for providing an extra length of yarn during cross weaving and it consists of a spring-loaded equalizer bar which is located between the whip roller and the

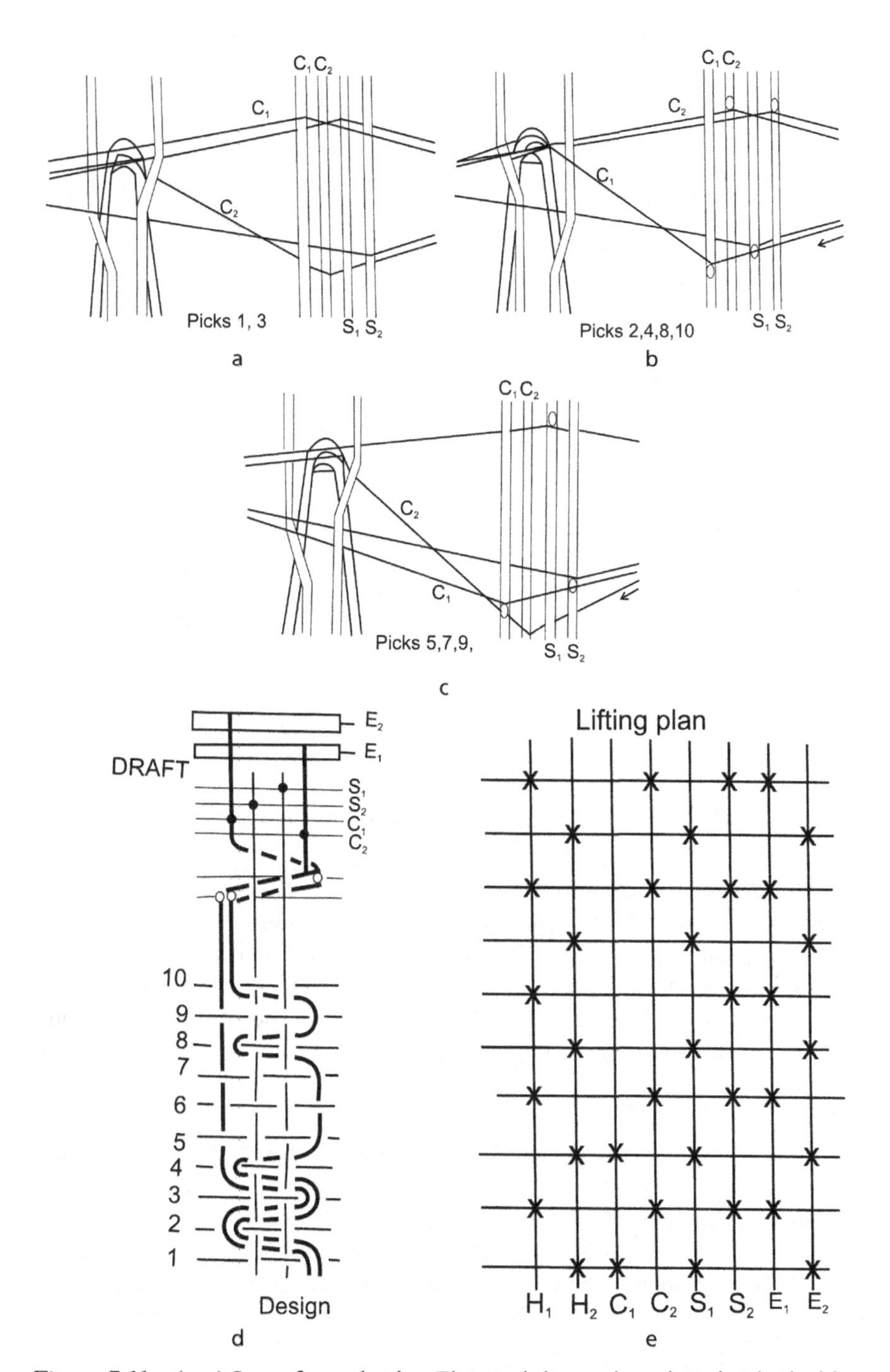

Figure 7.11 (a–e) Leno formed using Flat steel doup wires slotted at both sides

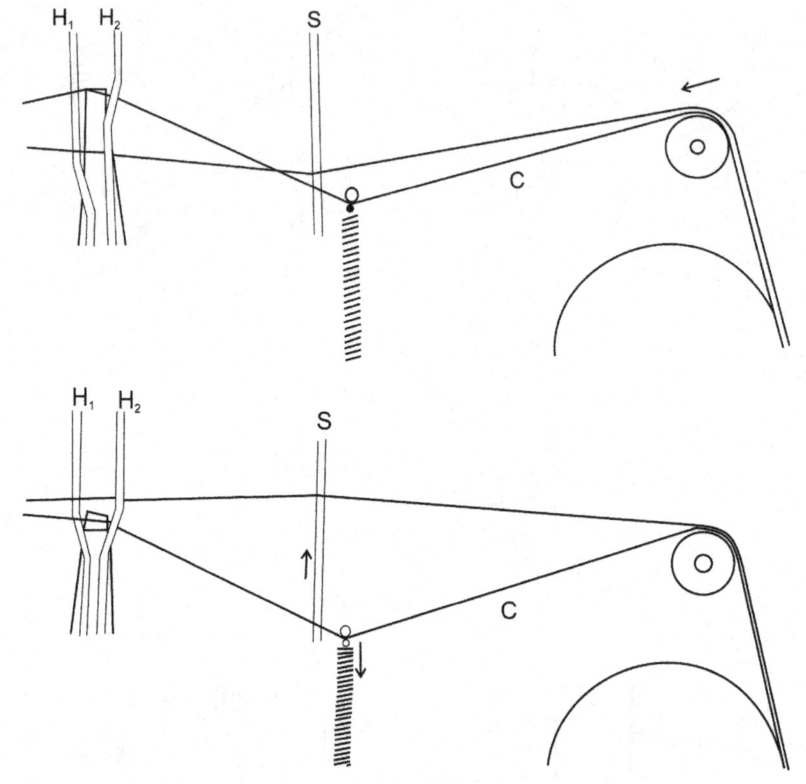

Figure 7.12 (a, b) Negative Easer motion

ground harness. Two springs will hold these bar down when the extra yarn is needed by the doup ends to make a crossed shed the doup ends will pull up on these easier bar overcoming the tension on the spring. This action allows the cross shed to be made. When the standard moves to make the open shed, the springs will return the easer bar to the down position and keep the doup ends light. Negative easer is particularly suitable for tappet looms (Fig. 7.12a and b).

7.9.1 Positive easing motion

Positively easing bars are normally mounted in the proximity of the backrest usually slightly above or below it. All the crossing ends are passed over the bar which at the required point in moved forwarded either by a dobby or by cam action to release a predetermined length of yarn. As the extra light is required only temporarily after each release which concedes with the crossed shed, the extra length is taken back either by a spring return action or by a cam (Fig. 7.13a to d).

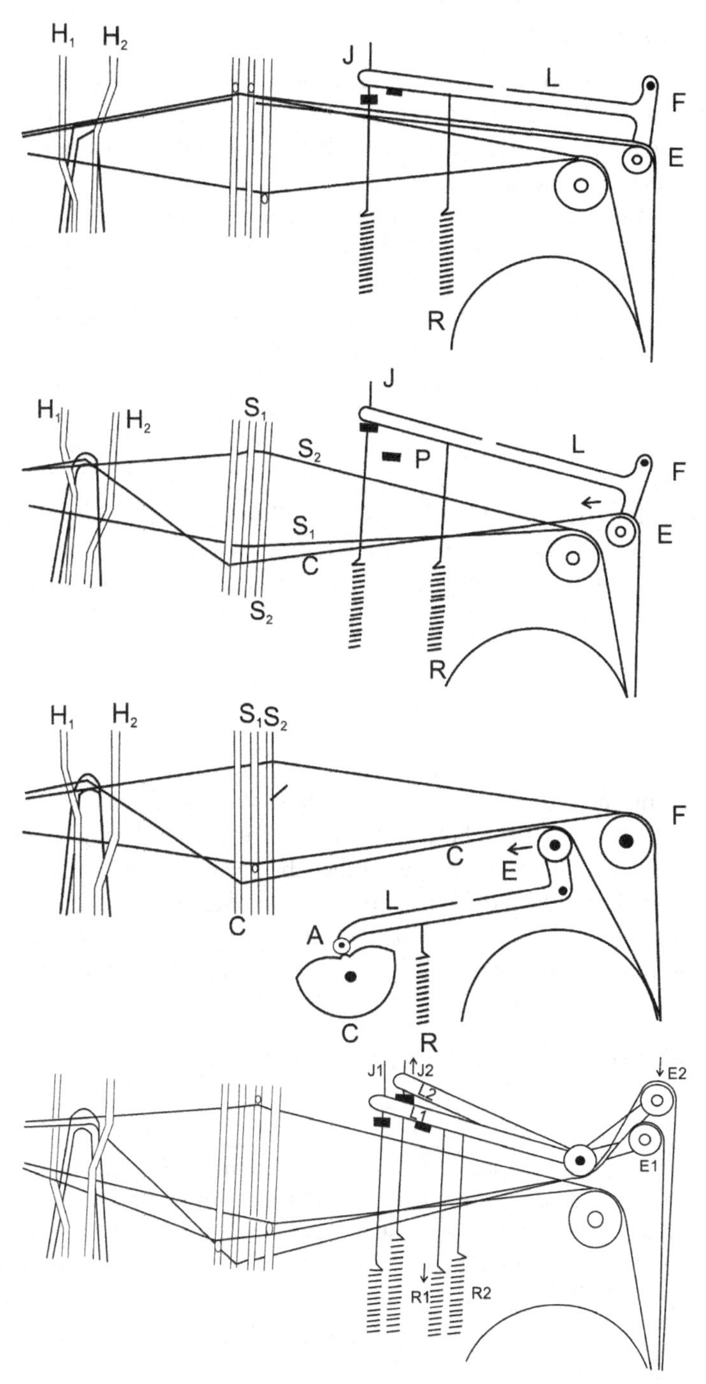

Figure 7.13 (a–d) Positive Easer Motion

7.9.2 Jumper motion

The doup and is upon the open shed and also upon the cross shed. The doup end moves crossed under the ground end, in order to do these. Since the ground ends are down on the open end, crossed shed, some means must be devised to raise it up, between picks, so that the doup end can cross under it. The shaker or jumper motion does just this. In short quick lifting motion, the ground ends are brought up slightly higher than the doup ends so that the doup ends may go under the ground ends in their traverse from open to crossed or crossed to open shed. The jumper motion is necessary only when the fabric is woven is open semi – open shed mechanisms. If closed shed mechanism is used the shaker is entirely unnecessary because all the ends after each shed are at the same level. The jumper can also be dispensed with when (a) simultaneous top and bottom douping assemblies are used and (b) when there are two or more standard ends in each leno group which weave plain. The half lift of the standard heald is obtained in various ways, either by an action or by a special attachment to a dobby which most makes now provide for the purpose. The attachments operate in a fixed manner and produce the shaking movement in between every pick whether it is required or not.

7.10 Yoke

The first or second standard will raise the doup needle. Gravity alone cannot be counted on the return of the doup needle to the down position. The doup needle has two slots in the bottom steel ribs are drawn through these slots and a yoke is attached to either end of the ribs one on each side of the loom. A spring is attached to the floor is connected to each yoke pulling the doup needle down either the first or second standard does not pull up on it. This is true for bottom doups. For top doups, the yoke will pull the doup needle up.

7.11 Differences between gauze and leno

S No	Gauze	Leno
1	The term gauze is applied to fabric woven on crossing principle of an extremely light, open, Himsy texture, and especially if produced from silk	The term leno is applied to fabrics woven on crossing principle with the weave applied decoratively to heavier texture of cotton and liven. They are further subdivided into net and leno brocade

S No	Gauze	Leno
2	Gauze weave is produced by causing one series of ends, termed doup incls to form more of less zig-zag or wary lines, whilst another series of warp threads, termed standard (regular) ends remain comparatively straight	Leno weave is developed by causing both standard and doup ends to bend equally as in leno brocade fabrics
3	Two beams are required	One bean itself can contain both the sets of warp
4	The plain gauze is the most popular weave. The gauze weave does not lend itself to much modification	Leno weaves one used in can junction with a host of other basic weaves, and a large no of decorative pattern is produced.

CHAPTER 8

Damask and brocade fabrics

8.1 Introduction

These fabrics are basically produced originally in rich silk material and are imitated with polyester, viscose rayon, and cotton. The fabric is composed of one series of warp and one series of weft with a maximum of thread set which is obtained by increased figuring capacity of jacquard. Even though they form single-layered structure, due to the complex method of production damasks are classified as compound fabrics. They are extensively used as ladies dress material (if produced in silk, viscose, polyester, acetate etc.), furnishing (if produced in cotton), draperies (surface-modified polyester). Damask is the technical term given to the certain distinct type of fabrics characterized by lustre using fibres cotton, wool, linen, silk. Satin and Sateen are the main weaves of damask. They are also characterized by the motif repeating on a large number of ends and picks which is mainly obtained by increased figuring capacity of jacquard. The term damask probably has its origin in the ornamental silk fabrics of Damascus which were elaborately woven in many colours and sometimes with Jari (fold or silver threads). At present, this denoted linen texture richly figured in weaving with various types of ornamental designs of flowers, fruits animals, and so forth. Damasks were manufactured in China, India, Persia, and Greece, but none could surpass Damascus in beauty and design and demand for her silken fabrics were so great that ultimately every silken fabric richly finished with the curious design, became commercially known as damask.

8.2 Classification of damask

Damasks are identified as true damask and one-sided damask depending on the use of type of weaves. In true damask figured fabric, weft sateen figure is formed upon a warp satin ground, or vice versa, and the structure is described as reversible. The term damask is however also applied to cloths in which the figured portion is developed in diverse ways upon a sateen or satin ground. The texture is then known as a one-sided damask. In addition to the reversing of the same weave or also two nearly similar weaves, for the development of the figure and ground portion of the fabric, in the manner just described and thereby obtaining a true counterchange of weaves which vertically results in a reversible fabric, those fabrics that are constructed on the true damask principle of weaving are further characterized by a more or less coarsely stepped margin of the figure and ground resulting from the method of controlling by means of the figuring harness, the warp threads in groups of two or more contiguous threads simultaneously and also of inserting two or more consecutive picks of weft during the operation of each pattern cord for the development of the design although both the warp threads and picks of weft each interweave independently in some definite order for the purpose of binding both the figure and ground position of the fabric in the manner described.

How exactly the damask motif repeats on a large number of threads even though the capacity of the jacquard used is small?

Yes, this is the correct and most opt question raised with respect to damask fabrics. It can be recalled that capacity of any jacquard can be increased either with or without making changes in the jacquard. And normally damasks are produced on a self-twilling jacquard (working is explained in the later part of the chapter). Increased picks are obtained by presenting the same card for more than two, three, or four picks as the case may be. The extreme fineness of texture that is usually obtained in these fabrics especially those of silk or fine linen containing sometimes as many as 400 threads, places them for beyond the range of the ordinary type of jacquard machines of reasonable capacity even for design repeating on the few inches only in the fabric. And to equip a loom with a sufficient number of ordinary machines would not only incur enormous expense but that course would also often times be quite impracticable.

In order, therefore, to overcome these practical difficulties and still enable true damask fabrics to be produced economically, there have been derived many very ingenious methods whereby a considerable number of warp threads may be controlled by a comparatively small number of hooks, needles, or both in the jacquard machine. Moreover, by using each pattern cord for several picks in succession, a pattern repeating on a large no of picks may be woven with one quarter, one third, or one half that numbers of pattern cords according to as each of the latter is used for four, three, or two picks, respectively. Thus, a

600 jacquard in which one needle control three ends and one cord of a set of 400 acted for three picks would produce a design repeating upon 1800 ends and 1200 picks. In addition, a design would need to be painted over an area of 6000×400 only thus achieving considerable saving in design preparation.

8.3 Self-twilling jacquard for damask production

This jacquard is available in fine and coarse pitch forms and working, on the whole, is similar to ordinary jacquard with the only difference in having additional hooks. Figure 8.1a shows the set up for a 400-hook jacquard with eight short rows. For each needle, A 3 hooks B are tied and thus the capacity is increased three times. However, selection of each of these hooks in the event

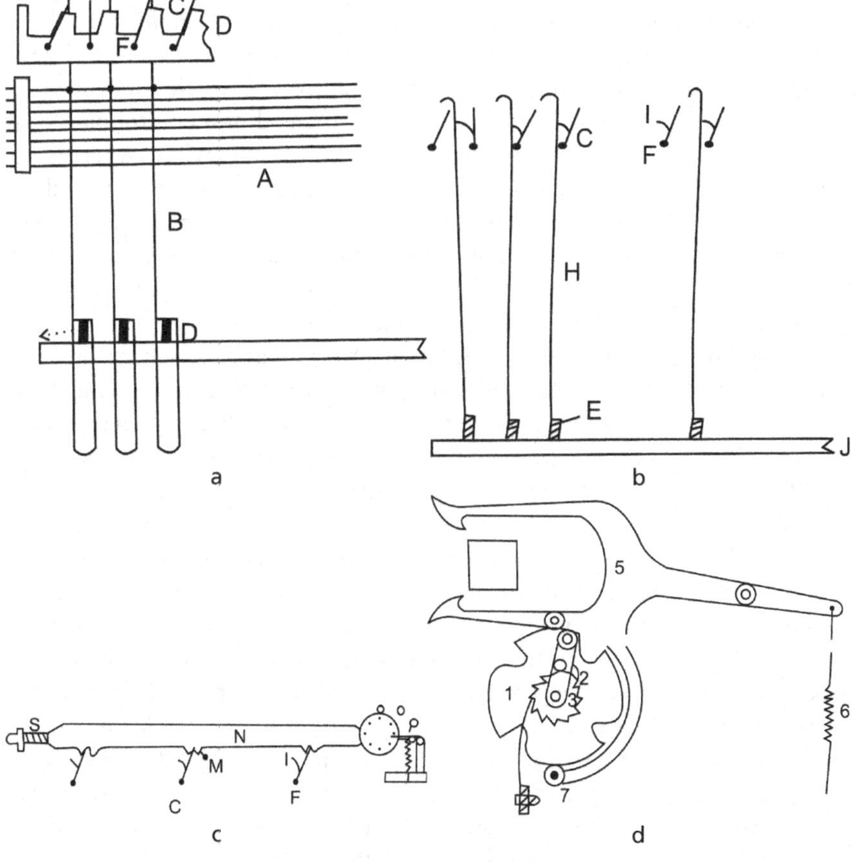

Figure 8.1 (a–d) Self twilling Jacquard

of rejection by the main jacquard is worth to note. Each hook at the bottom is bent and is resting on rods E which are supported by plate J. At the other end of the plate, the rod E accommodates twilling hooks H. The number of twilling hooks depends on the number of threads in the repeat of the binding weave used in damask. The bars E offer no obstruction to the lifting of the figuring hooks by the jacquard. The twilling hooks are provided for each long row of figuring hooks as shown in Figure 8.1b. The twilling hooks rest on J at their lowest position. The top portion of each figuring hook is turned towards card cylinder and are over the knives C in griffe G, but those of twilling hooks H are in the normal position held clear of the knives.

In their ordinary position, the swivelling knives C are inclined towards figuring hooks B in the usual way. Each knife C, however, is having a fulcrum at F in such a way that its upper edge is clear of the corresponding figuring hooks as shown by the knife 2 at Figures 8.1a and b. At the same time, the lip I of the knives which is in line with the twilling hooks pushes the latter over the preceding knife as shown by the hook 1 in Figure 8.1b. The knife C gets its rocking motion from a small revolving cylinder O in which projections or studs are fixed, each of the latter acting upon the end of a control bar N as shown in Figure 8.1c. The upper edge of each knife C fits within a recess M formed on the underside of a control bar N. As many bars are provided as there are threads in the repeat of the binding weave. Each bar N by means of recesses M control every fifth or every eighth knife, according to whether the biding weave repeats upon five or eight threads. The pressure of a stud upon O moves a control bar N (each of which is acted once per pick in every five or eight picks as the case may be) so that the knives to which the latter is connected assume the vertical position shown by knife 2 at Figure 8.1a. This knife is thus put out of engagement with the corresponding figuring row of hooks as shown in Figure 8.1a and into position for engaging the preceding row of twilling hooks as shown in Figure 8.1b. When the griffe rises on the following pick the figuring hooks in the long row 2 will thus be automatically left down. The twilling hooks 1 and every fifth or eighth twilling hooks; however, will be raised and lift up the corresponding bars E, and as each bar supports a long row of figuring hooks the rows 1 and so forth, will be automatically lifted. The arrangement causes one long row of figuring hooks to be left down and one long row to be raised to each repeat of the binding weave, quite independently of the figuring cards and each hook that is left down is next to a hook that is raised. That is, the jacquard lifts the ends in solid groups in forming the design, except that one in each repeat of the binding weave is left down through the action of the twilling motion, while the ends that are left down by the jacquard the same proportion is raised. A spiral spring S is used to return each twilling bar to its normal position after the pressure of the stud has been removed.

How to increase the figuring capacity of twilling jacquard?

Warp way figuring capacity

The figuring capacity of twilling jacquard mainly depends on the number of hooks per needle which in turn depends on the number of threads in the repeat of the binding weave. Normally, it may be two to four but sometime it may be two consecutive needle control two hooks and the third three, the fourth, two and so on The number of long rows of hooks is a multiple of the number of threads in binding weave repeat. For example, the given set-up in Figure 8.1a has a short row of eights needles and each needle control three hooks and thus it is most preferable to use eight-end binging weave. If however five-end binding weave is used then the number of long rows of hooks should be 20, 25, and so forth. For example, if it is 20, then the number of needles in a row of eight may control in succession 3,2,2,3,3,2,3 and 2 hooks.

Weft way figuring capacity

The number of picks in any jacquard controlled depends on the number of cards and it is impossible to accommodate more number of cards unless some arrangement like card cradle (reader is directed to refer 'Fancy Weaving Motions – K. T. Aswani' to get more details about card cradle and its arrangement) is ensured. But without increasing the number of cards, the number of picks per repeat of the design can be achieved in twilling jacquard. The logic is to feed the same card consecutively for two or three or four picks depending on the binding weave repeat size. This facility is provided in twilling jacquard and is shown in Figure 8.1d. The set-up consists of a ratchet wheel of specific size and can be changed based on the requirement accommodates a lever carrying a pulling pawl. The set-up also has a holding pawl to ensure only restricted movement of the ratchet. On the ratchet shaft, a cam with recess allows a bowl 4 or 7 and thus turning the cylinder for next pick. The bowl 4 or 7 will cause the cylinder movement either in forward or backward direction. As long as the bowls 4 or 7 are on the cam face no change will occur. But once the bowl 4 or 7 falls in the recess of the cam, it drags the lower catch to rotate the lantern of the cylinder. A cord connection 6 determines the shift in the position of the bowl either on the face of cam or recess of cam. By designing the cam suitably the number of picks per card can thus be controlled. For example, if the shape of the surface of the cam is such that it is possible to have two cards, three cards, two cards and so forth., successively, then weft figuring capacity can be increased accordingly. The cam rotation also depends on the number of teeth on ratchet. For example, let the ratchet be 24 teeth, then the cam is turned one 24th of a revolution each pick by the pawl. The design of the cam may be planned such that the raised portion of the cam 1 acts on bowl 4 or 7 for every two picks and the recess for one pick, then one card can represent three picks and thus the figuring capacity in weft direction is increased by three times.

8.3.1 Other equipment available for damask production

The main types of special systems employed to achieve the above results were the pressure harness, split, or banister harness, and the damask or self-twilling jacquards. Nowadays damasks are produced on ordinary high capacity fine pitch jacquards equipped with Jacquard soft wares operated through microcontrollers. Thanks to development in IT area as it has resulted in the production of online motifs fed from the keyboard or using what are known as computer-aided textile designs along with cardless or electronic jacquards. Although each of these several optional methods operate in a different manner in attaining the same result all are based upon the same principle, namely that of governing the warp threads in masse for the formation of the figure and individually for the purpose of bonding the warp and weft threads in some definite order of sequence.

8.3.1.1 Pressure harness

Here, a jacquard is used in combination with heald to move the same warp threads. The jacquard harness is for figuring purpose, while healed – the harness is for binding purpose. After passing thread, the mails H of the jacquard harness are groups of two, three, or more threads together, there are they passed separately through healds to form a straight draft. Heald eyes are made much longer than usual in order to permit of a full jacquard shed being formed within them the heald shafts, in turn, may be operated by tappets, dobby, or jacquard. When a jacquard is used to operate the heald shafts, each heald is controlled by two hooks which are connected by a cord supporting or small pulley and hook. The healds are suspended by cords passing from the pulley hooks and are pulled downwards, after being raised by means of weights or springs. By lifting two hooks that are connected or heald is raised to the 'top' position by lifting one hook only it is raised to 'neutral' position, and it remains at the bottom if both hooks are lifted down. The function of the jacquard figuring harness is to form the pattern by raising warp threads en masse for this figure and by leaning them down in masse for ground. The function of the heald is to produce binding weave throughout the entire fabric, by raising same of the warp threads that are left down and by keeping

Figure 8.2 (a–f) Development of Damask

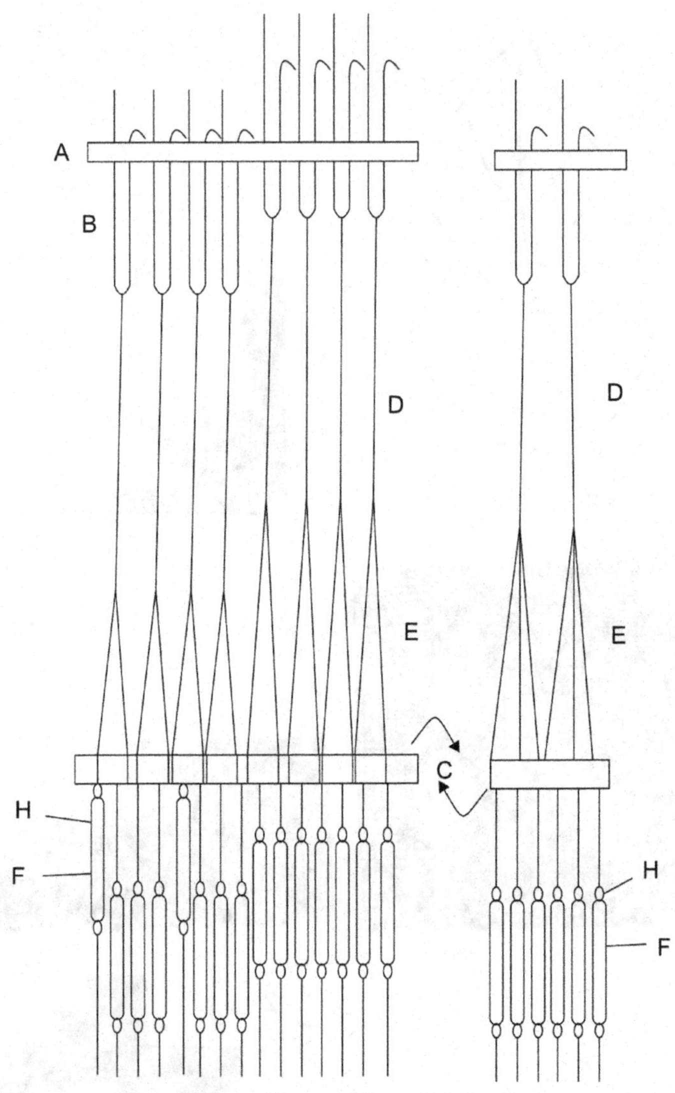

Figure 8.3 (a, b) Working of Split Harness

down others that are raised by the figuring harness. At least one heald must be raised and one left down for each pick, whilst the remainder occupies a neutral position in order to present any observation to the figure shed. The pattern cord is made to serve two or more picks of weft in succession by keeping the griffe knives of the jacquard only raised for such number of picks as are required to be inserted for each pattern cord, and then changing the cord for the next series of picks and are the griffe's must be raised every pick of weft, the changing of pattern cords must be accomplished quickly between the last pick of one pattern card and the first pick of the succeeding cord. The lifting cam governing the jacquard machine should be constructed to effect the change as easily and smoothly as possible otherwise it will cause considerable vibration to the harness and consequently bad weaving.

8.3.1.2 Bannister harness

In this system, additional harness is connected to each hook and these are operated by a special mechanism based on the binding weave. This system will lift a warp end which is lowered by regular jacquard by a different system. To understand the mechanism, it is necessary to assume a jacquard of a 400-hook capacity (Fig. 8.3a) with 50 × 8 arrangement. As per the card punch first four hooks are left down by jacquard and next or last four hooks are selected by jacquard. Based on the requirement additional harnesses are tied to each hook. Figure shows the increase in figuring capacity of jacquard is four times, and hence a total of 32 harnesses are observed. In other words, four additional harnesses are tied which carry a banister rod which is separately controlled by the same jacquard. The principle of bannister harness is that if the parent hook is lowered by jacquard, a warp end is lifted to form top line of the shed when it is acted on by bannister rod, in turn, may be lifted by one up and three down the principle for each pick.

The detailed working is explained as follows:

Figure 8.3a show the 400 hooks set-up in which first 4 hooks are lowered by the main jacquard and next 4 hooks are selected. Now attaché additionally two harness for each hook (that is 8 × 2 = 16 harness) A indicates the tug board, B the figuring hooks, C comber board, D harness, E additional harness, F Loop element, H banister rods. As each harness is tied with additional four harnesses, and all these additional harness are carrying banister rod. The rod is lifted in 1 up 3 down order (1/3 twill). As the rod is lifted, the warp end carried by the respective additional harness is lifted to form top line of the shed even though the main hook is lowered. It can be observed that for the first pick, first of the four banister rod is lifted and on the second pick, the second rod of each group of four is lifted and so on. The rod is not operated for those ends which are lifted by jacquard itself.

Two different positions of the Bannister rod can be possible.

1. At the top of the loop of each additional harness – owing to the lifting of banister rod in a specific order (here, it is 1 up and 3 down).

2. At the centre of the loop of each additional harness – bannister rod is not operated.
3. At the centre of the loop when the hook is selected by jacquard (all harnesses are lifted to form the shed).

The order of lifting of the banister rod do depend on the binding weave such as twill and the number of additional harness tied to each parent harness.

8.4 Methods of preparing damask designs

Designing is very simple requires very little technical skill. The structural feature of the design is dependent on mechanism governing warp threads. Following are the steps of designing an all over design for damask.

1. Consider the motif or figure to be produced as damask
2. Develop the motif or figure profile on the point paper
3. Select the weaves for various portions of the figure
4. Calculate the count of the design paper based on the ends/cm and picks/cm
5. Based on the number of threads in the binding weave, make necessary arrangements for increasing the figuring capacity
6. Based on the repeat size of the binding weave, develop the figure on the point paper which is a sateen paper
7. Three types of marks are observed in the final developed design. namely: solid marks indicate the warp up, the blanks in the figure represent ends that are left down due to knives missing the figuring hooks and crosses in the ground indicate the lifts produced by twilling hooks
8. Count of design paper for designing in twilling jacquard:

   Ends/inch: picks/inch

   Hooks/inch: pick/cord

Example is shown in Figure 8.2a – motif under consideration, b – developing the profile (a part of the motif is shown for understanding purpose), c – developing the figure on sateen paper, d – developing the figure based on the repeat size of binding weave (in the example, the repeat size of the binding weave is eight – ends and are controlled as 3,3,2).

Card Cutting instructions: Cut all the marks except blanks.

8.5 Types of damasks

Damask fabrics comprise of two chief modification, namely:

- Simple damask containing only one series of each of warp and weft threads as exemplifies in worsted and linen table covers table linen, light curtains, and hangings.
- Compound damasks containing one series of warp threads and two series of weft threads namely 'face' and 'back' picks, respectively, as exemplifies in some varieties of furniture upholstering fabrics, curtains, and hangings.

8.5.1 Compound damask fabrics

Following are the types

1. a) Those produce with an additional series of weft threads

   b) Those constructed with two distinct series both of warp and weft with the object of producing textures of greater strength bulk and weight.
2. In both (a) and (b) extra series of weft constitutes backing picks only with the primary object of imparting to the fabrics additional firmness and stability. In (b) extra series of warps in meant only for bonding purposes – they play no part in development of design.

8.6 Commercial damask fabrics (arrangement of figures in jacquard weaving)

In market majority of damask fabrics are being produced in decentralized sector like handloom items.

Motifs in these fabrics are found to be arranged on various basis as noted below

1. Either on full drop, half drop, one third drop, or one-fourth drop, and so forth.
2. Motifs may be on the basis of floral, animal, plant, geometrical, drop reverse base, diamond base, ozee base, ozee stripe base, freehand base, and so forth.
3. Motifs are also arranged on cross-border or running border or butta or all over type.

Fabrics are produced in wider width such as 58", 64", or 70", and so forth and in 100% cotton, or 100% cotton weft and viscose warp or any other combination. The advantage of viscose warp and Cotton weft (10 's–14's) is that figure is prominently brought on the surface and motif will be lustrous. The latest Damsak are being produced using computer-aided textile design as today in India at least half a dozen companies are versatile with textile jacquard designs along with complete calculations. In this regard, it is worth mentioning that pictures of great leaders or nature or epic are produced singly (i.e. only one large motif is observed) in the fabric.

8.7 Brocades

The term Brocade signifies a class of heavy richly decorated shuttle woven silk/cotton/man-made fabric ornamented with raised figures formed by extra threads or by embroidery. According to Nisbeth, the term can however be employed as a generic term virtually comprising all varieties of woven fabrics of simple texture consisting of only one series each of warp and weft threads and as distinct from compound types of fabrics of more complex construction. It is more limited application however, the term Brocade refers to more particularly to the lighter and medium weight texture of silk, linen, and cotton fabrics of simple structure and embellished by more or less elaborate jacquard figuring which may be developed by displaying either warp or weft only or both series of threads, more or less freely upon a ground texture usually consisting either of plain calico or tabby weave or other simple and neutral weave such as the smaller twill, satin, and matt weaves according to the particulars effect desired in the fabric.

Thus simple brocades can be produced in three ways:

1. Weft figuring
2. Warp figuring
3. Warp and weft figuring

8.7.1 Developing applied design for brocade fabrics

Following are the steps

1. Sketch design on squared or point paper.
2. Margin is indicated by single-dotted line. Alternate warp threads raised only on alternate picks. Intermediate warp threads raised only on intermediate pick. Therefore, two or more continuous warp threads

are never raised at the same time, nor is a warp thread ever raised for two or more consecutive picks/weft and so forth.

3. Outline is complete as in (2) and then braiding of floating weft figures with elementary weaves to prevent too long floats in dine.
4. Binding weaves should have single binding points.

8.8 Comparison of damask and brocade

S No	Damask	Brocade
1	Not reversible figuring effect on one side only	Reversible fabric
2	Any of the basic weaves like plain, satin twill, matt diamond and so forth may be used for the binding purpose	In true damask only satin satern are used. However, satin/satern and twill or twill and trill combination can be employed. No other weaves
3	Figuring capacity is increased by employing splits shed method of lying on a DL DV jacquard	Spenial system such as ____ harness, banmister harness, self-twilling jacquard and so forth necessary.
4	Preparation of design very easy. No need to indicate binding points as it is automatically accomplished by the jacquard equipped with special arrangement	Preparation of design is complicated and laborious, especially when a number of basic weaves are used for the different portions of the design

CHAPTER 9

Practical aspects of fabric analysis

9.1 Introduction

Fabric analysis is an important activity in the production of a new design for garment or apparel industry. Indeed fashion changes every minute in the world market and it is necessary to plan accordingly. Responding to the changes in the market it is this activity prepares the designer to plan for the new strategies. In fabric analysis, the fabric under production will be designed based on the specific end use. Fabric analysis gives all details from sample data to manufacturing data.

Following are the points to be kept in mind while analyzing a single-layered fabric

1. Check whether the fabric is grey or finished or stitched to size or hemmed to size
2. Identify the class to which it belongs
3. Identify the face and back
4. Check is it dyed or printed at one side or double side, if printed at one side, the analysis will be resumed from face side only
5. Measure the dimensions such as length, the width of fabric randomly at various points (select at least five observations), and report the average
6. Measure the selvedge width and count the number of selvedge ends
7. Identify the weave of selvedge
8. Mark the warp direction on fabric by an arrow for all future warp parameter measurements.
9. If in a piece of fabric where no selvedge is available, proceed as follows to locate the warp in the piece of fabric

- Ravel a small length of the sample in any direction and count the number of ends with either pick glass or densimeter. If the number of ends are more, the direction is marked as warp
- If in a piece of fabric threads/inch is the same in both the directions, then check for reed marks in the cloth and that direction is marked as warp

10. In the long piece of fabric, using densimeter measure the ends/inch and picks/inch at random places at least for five times and report the average
11. Measure the count of warp and weft using Beesley balance. Repeat the experiment for at least five times and report the average value
12. Measure the crimp of warp and weft using crimp tester
13. Ravel the threads and analysed the working of an end passing through different picks and mark the design on the point paper
14. After locating the repeat, apply the rules of drawing and find the number of heald shafts required and compare with the total number of ends in the fabric considering the full width of the fabric. Sometimes as per design only two heald shafts are needed but the number of ends may be more and in such cases, number of heald shafts to be calculated on the number of ends
15. Make the calculations for manufacturing data

9.2 Fabric analysis – a tool for fabric engineering

9.2.1 Case – I: Single-layered structure

Every weaving factory or unit will be harping for new designs or new constructions through weaving. New design aspects of a fabric can be planned even in chemical processing section or unit of a composite unit or a decentralized unit. Everybody's eye is now on fashion – a fast-changing tradition. (The role of films and cinematography is to be mentioned here.) Specially one should note that today most of the heroes in any film will wear jeans and a number of varieties of jeans are made available today. How is this possible? Who is behind the show? How exactly a new design or fashion is produced? Answers to all these questions lie with the functioning of 'fabric analysis and design' wing in any textile industry.

9.2.2 Role of R&D department

It is the SQC (Statistical Quality Control) and QAD (Quality Assurance Department) carryout the thrust for new designs. In this process, there are two ways. One way includes designing a totally new fabric through fabric engineering based on the end-base requirement and planning of manufacturing facilities accordingly. The second way considers the new design which is popularly accepted by consumer and carryout fabric analysis in a deep style. Planning for the manufacture of data based on the fabric analysis will be the next step. Based on the modified parameters such as count of warp and weft, the width of finished fabric, type of finish to be given, and so forth. We find today the second approach is followed and found widely accepted in India.

9.2.3 Need for fabric analysis

Following are the objectives of fabric analysis:

1. To determine the particulars of warp and weft such as count, twist, diameter, blend composition, type of fibre, and so forth.
2. To determine the details of fabric such as cover factor, fabric quality index. Fabric weight factor, fabric characteristics, and special features based on specific end use.
3. To determine the manufacturing details such as loom type and loom data (total number of heald eyes/shaft, pick wheel, reed count, warp length, the width in reed, and so forth).
4. To plan for specific chemical processing of woven fabric by applying special finishes.
5. To study the formability and tailorability of the proposed new fabric.
6. To examine the 'drape' and 'drop' behaviour of woven garment.
7. To plan the large or mass customization of the new design by integrating the facilities of CATD (Computer Aided Textile Design) (computer-aided woven fabric garmenting using cut planning) auto laying, automated knitting, sewing (special finishing if any), care labelling and packing and shipping.
8. To test the pulse of the consumer or market through the launch of fashion shows.
9. To plan for regular production with corrections if necessary based on the feedback from consumer following market research.

9.2.4 Procedure for fabric analysis and design

Three stages are involved

1. To determine the fabric analysis particulars as explained in the next phases.
2. To workout, the manufacturing facilities required based on the information from the above step.
3. To get various combinations of colour and weave match through textile software such as texture mapping, jacquard, or dobby design or print software.

Fabric analysis – stage I

First, it is necessary to examine about fabric condition, that is, grey or dyed or printed. It is grey or monocoloured dyed analysis can be started either on the face or back side of fabric under consideration. However, if it is single-sided printed, the analysis will be on the face side only. The above discussion shall hold good only in plain fabric (Y_1) case. But in the fabric where the float is present, it is recommended to analyse in face side (warp over weft) only.

Following are the steps:

- Check for the selvedges and measure the width and mark the direction of warp direction by ↑ mark.
- If it is a sample of smaller length, weigh the fabric and note the weight in grams (replicate the step five times with new random samples)
- If it is from long (Than) roll, cut 2 meters and weigh (choose five random samples).
- Measure the length and width at five different places and note average length and width.
- Compute arial density using weight (grams)/area (m^2).
- This is referred to as direct method for determining fabric weight.
- Of course, it should be noted very clearly were that any finished fabric will have respective finishing agent or a grey fabric will have certain percentage of size and due to this may not match with direct method unless the size percentage is added.
- Cut 10 cm × 10 cm fabric sample by taking care in avoiding weaving or processing defects. Weight in a quadrant balance or physical balance and compute weight (g/m^2)=mass/area.

- This is known as 'template method'. Repeat this experiment for five replications by considering random samples.
- Proceed to weight determination by emperical formulae

$$W = 0.6857\left[\frac{n_1}{N_1}(1+C_1)+\frac{n_2}{N_2}(1+C_2)\right]$$

where W = weight in oz/sq.yd

n_1 = ends/inch

N_1 = warp count (english)

C_1 = warp crimp (%)

n_2 = picks/inch

N_2 = weft count (english)

C_2 = weft crimp (%)

Or

$$W = 0.1\ [n_1N_1\ (1 + C_1) + n_2N_2\ (1 + C_2)]$$

Where

W = weight in g/m^2

n_1 = ends/cm

N_1 = Tex warp count

n_2 = picks/cm

N_2 = Tex weft count

C_1 & C_2 – Warp and weft crimp (%)

Whether it is 'w' in oz/yd^2 or g/m^2. It is necessary to determine first the direction of warp as the other thread will be weft

How to determine warp direction in a fabric?

Case (i) – When the fabric is available in full width, it is easy to mark warp as it will be running parallel to the selvedge

Case (ii) – where selvedge is not available

Approach I: ravell the threads in one direction and using pick glass, count the threads/inch. Now ravel in the other direction and count threads/inch. The direction in which threads/inch is more will be warp (this is based on the fact that the majority of fabric are warp faced in nature ($n_1 > n_2$ or $w_1/w_2 = 1.01–3.01$)

Approach II: Use densimeter and find the direction in which the density of threads/inch is more and note it as 'warp'. But it is to be noted that in some cases where the fabric is dyed in heavy shades/colour, densimeter will not be useful.

Case (iii) – Check for reed marks in fabric and mark the direction of warp parallel to reed mark

Step I

As described above, determine ends and picks/unit area either by using pick glass or densimeter (note: densimeters are available in different ranges, it is necessary to select a right type by observing the threads closeness visually). Repeat the experiment five times by selecting random places and record it as n_1 and n_2 (convert the readings into epc or ppc by dividing from 2.54).

Step II

Using Bushy balance, determine the count of warp and weft or ravel a thread (warp/weft) of known length and weight it, compute c = l/w or w/l as the case may be. In any case, repeat the experiment five times and record the average values.

Step III

Measure the crimp of warp and weft as explained below:

Approach I: Through manually

Ravell a known length of warp or weft from the fabric using dissecting needles by taking care without disturbing the crimp. Place the known length thread on a scale and by holding strongly one end, stretch the other end to a maximum extent so that the waviness or crimp is removed. Note this length as l_2 and known length as l_1. Crimp is calculated as

$$l_2 - l_1/l_1 \times 100$$

Approach II

- The crimp can be measured using the instrument. In any of the case, five observations are recorded and the average value is reported. Compare the weights by all the three methods (note: the difference should be ± 2)
- Compute warp cover $K_1 = n_1/\sqrt{N_1}$
- Compute weft cover $K_2 = n_2/\sqrt{N_2}$

- Compute fabric cover $K = K_1 + K_2 - \frac{K_1 K_2}{28}$
- Measure yarn (warp and weft) diameter using Shirley thickness gauge
- Also, compute yarn diameter using pierce formulae
- Examine the type of yarn (single and double) and twist direction
- Measure tpi using respective twist testers
- Compute $\text{TM} = \frac{tpi}{\sqrt{count}}$ for warp and weft for the indirect system or $tpi \times \sqrt{tex}$ count for the direct system
- Examine the type of fibre (natural/regenerated/synthetic)
- If it is a blended yarn, determine the blend % by suitable methods
- Note the special features in text form highlighting following:
  a) Type of fabric – printed/dyed/finished
  b) Type of yarns used – coarse, medium, fine
  c) Type of yarns used – single/double/cable
  d) About cover factor – open set, medium set, close set
  e) About nature of fabric – warp faced ($n_1 > n_2$), weft faced ($n_2 > n_1$) or nearly equifaced ($n_1 = n_2$)
  f) About crimp levels, that is, is $c_1 > c_2$ or vice-versa
  g) About the type of fabric and its blend
  h) Note about TM
  i) Its end use and probable width of sale
  j) Suitable loom equipment

Computation of manufacturing data – stage II

Compute the following:

1. Total number of ends = epi × body width (in) + number of selvedges

   Or TE = epc × body width (cm) + number of selvedges
2. Number of heald shafts
   i. According to design – analyse the fabric on face and mark the repeat. Extracting the repeat, draw draft, denting and lifting plans

ii. According to calculations – generally 1000–1500 ends are drawn/shaft.

3. Number of heald eyes/shaft

$$\text{For body} = \frac{\text{Total number of Body ends}}{\text{Total number of Heald shafts} \times \text{Drawing order}}$$

$$\text{For selvedge} = \frac{\text{Total number of Selvedge ends}}{\text{Total number of Heald shafts} \times \text{Drawing order}}$$

For selvedges drawing order (generally) is 2 ends/eye

4. Width in reed (cm) = cloth width (cm) $(1 + C_2)$
5. Warp length (m) = Fabric length (m) $(1 + C_2)$
6. Stock port reed count = $\frac{n_1}{1+C_2}$ (Always even number and in case of fractions nearest even number)
7. Pick wheel = $\frac{n_1}{1+C_2}$
8. Warp weight factor $w_1 = n_1/N_1$

   Weft weight factor $w_2 = n_2/N_2$

   Fabric weight factor $w = w_1 + w_2$
9. Fabric Quality Index (for plain weave) $K_1 + K_2/32.2$

Selection of loom equipment details – stage III

1. Type of shedding device: Dobby/Tappet/Jacquard
2. If Dobby, number of jacks required =

   If Jacquard, size required =
3. If Jacquard, total number of cards required =

   Harness tie up suggested =

   Card cutting and casting out details =

Other details – stage IV

1. Fabric name
2. Fabric type (weave and sort) Eg. plain, long cloth
3. Fabric condition: grey/dyed/printed/finished

4. Variety: medium A or B
5. Sort number
6. Manufacturers name:
7. Tex.mark.number

9.3 Significance of each parameter in fabric analysis

While undertaking fabric analysis one will notice a number of features regarding the nature of yarn, type of twists, level of twists etc. Following paragraphs give the reader a basic idea about parameter significance in fabric designing.

1. **About fabric and its dimensions**

   Fabrics may be simple or single layered (one series of warp + one series of weft) or semicompound (bed ford cord/pique/distorted thread effect) or complex fabric (more than one series of warp and weft).

   Here, the discussion is limited to simple fabrics.

2. **Width**

   Based on width of fabric, one may identify as

   a) Narrow width
   b) Medium width and
   c) Wider width fabrics and accordingly the specific end use is explored

9.3.1 Selvedge and type

Selvedge type like conventional, tuck-in, leno, and so forth is also considered. The width of selvedge is measured. Normally, it is 0.5″ on a side, but today selvedges are 0.75″ and an additional selvedge for company name or type of sort and so forth is also noted. Selvedge weave is examined and posted on graph paper. Normally, selvedges are produced with 2/1 rib weave.

9.3.2 Threads/unit space

Different situations may be noted. If $n_1 = n_2$, give equifaced fabrics. But in weaving, it is well-known fact that reed and pick have different values and even though, if they are same, crimp of warp and weft differ and hence one has to manipulate reed and pick with the knowledge of crimp to get $n_1 = n_2$.

In most of the cases, constructions are warp function in nature, that is, $n_1 > n_2$ and thus weaving of fabrics does not pose problem unlike $n_2 > n_1$.

In some special cases such as casement, PPI (Picks per inch) is more than EPI (Ends per inch) giving weft faced fabrics. In such cases, we observe bumping conditions during weaving (pick wheel>reed count). More details can be had from 'principle of weaving' by ATC Robinson and marks.

Thus, it is interesting in fabric analysis to note any one of the conditions like $n_1 = n_2$ or $n_1 > n_2$ or $n_2 > n_1$ and the constructions will depend on specific end use. It also mentioned here that thread density governs properties such as air permeability tensile, tear, and so forth.

One should note that thread spacing in relation to yarn count; that is, finer the count, higher is the thread spacing and accordingly particular type of densimeter to be used.

Higher the threads/inch, higher is the fabric cover. For example in cotton or nylon typewriter ribbon clothing, a maximum cover is desired.

9.3.3 Yarn count

Yarn count is the specification for yarns. It states about the nature of the yarn. Yarn count do depend on the type of fibre and nature of the fibre. Yarn count is selected based on the specific end use of fabric. Depending on the yarn count to be spun, tpi depend in yarn formation stage. Based on the count, parameters such as reed count, heald count, adjustments for shuttle box, race board, type of weft fork and type of weft fork mechanism, type of let-off, type of emery fillet count depends. Yarn count also a contributing factor for fabric cover and hence in case of pure silk fabrics made from 20/22 D, usage of 28 s reed and 90 pick can be observed. On the other hand, based on nature of fibre such as regenerated or synthetic and corresponding denier, thread density in fabrics like sateens is decided.

Selection of dyes, nature, and type of dyes also depend on nature and type of fibre and yarn count. Thus, the yarn count has a significant effect on fabric texture and its properties.

Yarn count also governs fabric properties such as tensile strength, tear strength, ballistic strength, abrasion resistance, air resistance and so forth. Most important properties which are controlled by yarn count are fabric bending, crease recovery, wrinkle recovery, and fabric drape. Needless to mention here about the surface properties such as friction and other properties like compression do depend on yarn count.

Yarn count also governs the production in kg/shift or yarn content in grams/ spindle in ring spinning. Coarser counts have lower doft time as compared to finer counts. Specially, if wage payment is based on piece rate system, operators do prefer coarser counts.

It is also noted here that applications such as casual wear, rough wear, sportswear, royal wear call for the selection of proper yarn count.

In the majority of the cases warp and weft counts are equal so that one out of three factors (count, thread spacing and crimp) are maintained constant so as to achieve approximately square set fabrics.

9.3.4 Fabric cover

The cover factor is defined as he extent of coverage of yarns in the fabric. The maximum cover can be a situation in which the yarns touch closely with the adjustment yarns without air spaces.

Cover plays an important role in fabric texture and application of specific end use. For example, in the case of window curtains, openness is most important and hence open set constructions are preferred. This condition can be had either from coarse (window curtain) or fine and medium counts (some of the shirting construction with leno weave). Another interesting example is that of sail clothes made in 2/2 matt weave. It is well-known fact that matt weave give open constructions.

Based on cover factors plain fabrics are classified into open set (with cover factors 10–22) and close set (with cover factors 22–35). Needless to mention here that yarn diameter or yarn count is an influencing factor for fractional cover (d/p) cover factor is used to compare relative closeness of different fabrics. Parameter like the type of weave cannot be ignored here as some type of weave like mock leno, matt weave where thread work in groups or threads in the group are drawn in one dent have to influence on the cover. Best examples for cover factor can be observed from jute gunny bags (of course today PP is used in leno weave for vegetable onions packaging or transportation) as they are made faith different covers and for onion it is fully open, for cotton portation (earlier practice) moderately open, for rice package nearly close and for sugar complete close). But here a mention is made with respect to sweaters of wool. Even though woollen yarns are used, the fabrics are medium covered and warmth ness is due to wool and the cover factor is the secondary factor for fabric weaving.

9.4 Fabric analysis sheet: Simple structures

Face			Back	
Class of Fabric			Commercial name	
Sample analysis data			Manufacturing data	
Parameter	Warp	Weft	Parameter	
Thread density ($n_1 \times n_2$)			Total number of ends	
Count ($N_1 \times N_2$)			Total number of heald shafts:	
Crimp ($\%c_1 \times \%c_2$)			(i) According to calculations	
Cover factor ($k_1 \times k_2$)			(ii) According to design	
Fabric cover factor ($k_1 + k_2 = K$)			Total number of heald shafts	
Yarn diameter			Body......... Selvedge..........	
i) Gauge (mm)			Reed count	
ii) Formula ($1/28\sqrt{N}$)			Ends/inch in reed	
Twist (*TPI*)			Width of reed	
Twist multiplier (*TM*)			Warp length	
Type of fibre			Pick wheel	
Type of yarn			Warp weight factor	
Blend composition			Weft weight factor	
Fabric weight			Fabric weight factor	
i) Direct method			Fabric quality index	
ii) Template method			Fabric name	
iii) Empirical method			Fabric quality	
Special features			Variety/sort	
			Manufacturer name	
			Fabric condition	

9.5 Advanced fabric analysis sheet

Face			Back
Class of fabric:			Commercial name:
Type of motif:			End use:
Sample analysis data			Manufacturing data
1. Thread density ($n_1 \times n_2$)	Warp	Weft	
Ground/pile			
Ends/inch and picks/inch			a. RTP
Face/back			
Ends/inch and picks/inch			b. Total number of ends
2. Count of yarn ($N_1 \times N_2$)			c. Denting order
Ground/pile			
English count (N_e)			d. Reed count
Face/back			
English count (N_e)			e. Number of beams required
			f. Yarn preparatory requirements
3. Crimp ($C_1\% \times C_2\%$)			Beam/sectional
Ground/pile			
Face/back			g. Loom equipment
			i. Plain/dobby/jacquard/terry
4. Direction of twist			
Ground/pile			ii. No. of shafts/jacks
Face/back			
			iii. Size and type of jacquard selected
5. Type of fibre			
Ground/pile			iv. Total number of hooks employed
Face/back			
			v. Number of cards to be punched
6. Fabric features			vi. Card cutting and casting out inst.

Appendix

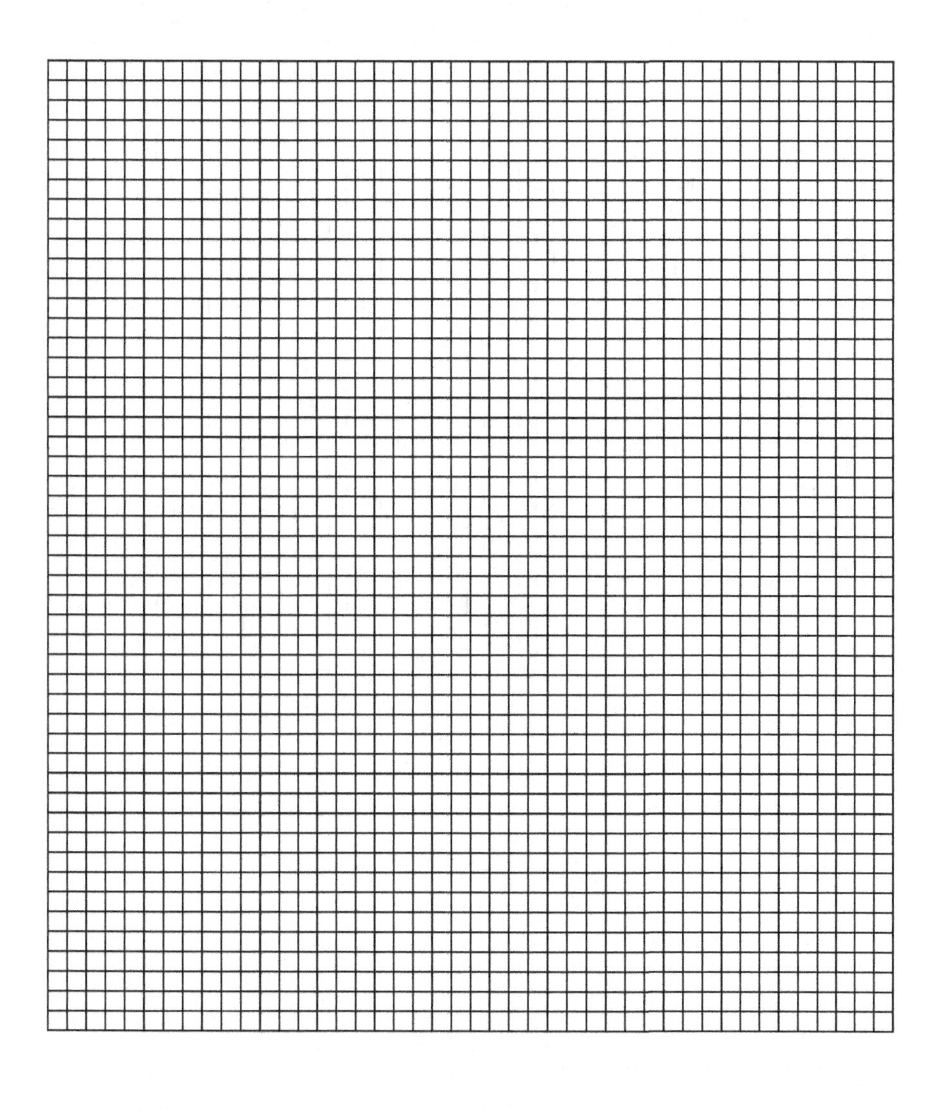

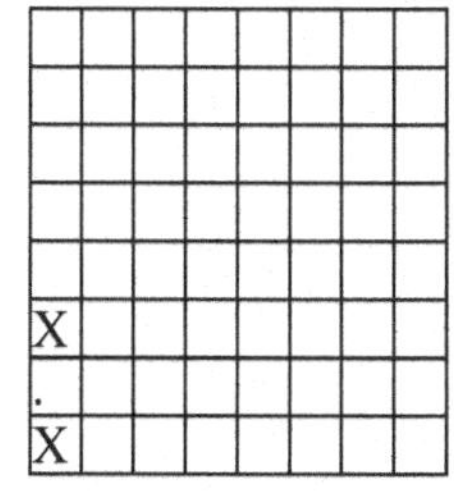
X
X

Index

Printed in the United States
by Baker & Taylor Publisher Services